Sustainable Land Management: Issues, Problems and Prospects

Sustainable Land Management: Issues, Problems and Prospects

Editors
Prasanta Kumar Mishra
Narinder Kumar Sharma
Debashis Mandal
Gopal Kumar
Rajesh Kaushal
Anup Das
Raman Jeet Singh

SSPH

SATISH SERIAL PUBLISHING HOUSE
403, Express Tower, Commercial Complex
Azadpur, Delhi-110033 (India)
Phone : 011-27672852, Fax : 91-11-27672046
e-mail : info@satishserial.com, hkjain1975@yahoo.com
Website : www.satishserial.com

Published by :

SATISH SERIAL PUBLISHING HOUSE
403, Express Tower, Commercial Complex, Azadpur, Delhi-110033 (INDIA)
Phone : 011-27672852 Fax : 91-11-27672046
E-mail : info@satishserial.com, hkjain1975@yahoo.com

ISBN 978-93-88020-46-6

Composed, Designed & Printed in India

About the Editors

Dr. Prasanta Kumar Mishra, Consultant, IDC, ICRISAT and the Ex-Director of ICAR- Indian Institute of Soil & Water Conservation, Dehradun has a distinguished academic and service career as a student, field engineer and a researcher in Soil and Water Conservation Engineering spanning over 35 years.A graduate from OUAT in Agricultural Engineering and M. Tech. from IIT, Kharagpur in Soil & Water Conservation Engineering, he studied at University of Karlsruhe, Germany for his Ph.D. under DAAD fellowship from 1986-1988. He was trained in Professional Development Program in UK and in the field of GIS and Hydrologic Modelling in USA and Australia. He extensively worked on soil and water conservation measures by developing unique facilities like rainfall simulator and tilting hydraulic flume for resource loss estimation and compiled useful books on soil conservation and rainfed technologies. He worked in various capacities at IIT, Kharagpur, Govt. of Odisha, ICAR-CRIDA and AICRPDA as Project Coordinator and ICAR-IISWC. His major contributions are in the area of drought management, documentation of ITKs relating to Soil and Water Conservation, Common Pool Resources (CPR) management, development of research watersheds, compilation of watershed and rainfed technologies, and climate change research initiatives through various DFID, ACIAR, World Bank, Indo-Austria, DAAD, ICAR and DST projects. He developed unique Drought Management Software for Andhra Pradesh and compiled a book for ICAR. He invented low discharge and low cost Baffle-sluice irrigation module for volumetric supply of irrigation water. He visited USA, Australia, Germany, UK, The Netherlands, Austria, Italy and Nepal on various research and training assignments. Dr. Mishra is the recipient of a number of awards which include DAAD fellowship from Germany, best undergraduate student award by ICAR, New Delhi, Student Award and Fellowship from Indian Society of Agricultural Engineers (ISAE), New Delhi, Jawaharlal Nehru Award of ICAR, New Delhi for Best Ph.D. work in Engineering and Technology, Fellowship Award and Dr. K.G. Tejwani Award for Management of Natural Resources of Indian Association of Soil & Water Conservationists (IASWC), Dehradun, SCSI "Gold Medal" and Fellow Award by Soil Conservation Society of India (SCSI), New Delhi, Vasant Rao Naik Award (Twice), Team Award and Nanaji Deshmukh ICAR Team Award of ICAR, Ground Water Augmentation Award (as a team member) from MoWR, GoI and Fellowship (FIE) and Eminent Engineer Award from Institution of Engineers (India). He has about 300 publications to his credit including 140 research papers in

National and International journals of repute, books, book chapters, papers in national and international proceedings, popular articles, bulletins and reports and short communications.

Dr. Narinder Kumar Sharma, specialized in Conservation Agronomy, Agroforestry and Watershed Management and has about 33 years of research experience in the areas of resource conservation and management; agronomy/agroforestry for conservation, production and environmental enhancement; soil fertility/land use/watershed management/Farming system research, Utilization of degraded lands; upliftment/social service to the society and sustaining the Himalayan Agro-Ecosystem in context of changing climate. He worked on several national and international research projects funded by agencies like ICAR, NATP, DST, NAIP, USUAID, MoA and GoI. Presently, he is working as Head, Soil Science & Agronomy at IISWC, Dehradun.

Dr. Debashis Mandal born on 30th September, 1969 in West Bengal, obtained B.Sc. (Ag.) and M.Sc. (Agri.) in Soil Science Vishva-Bharati, Shantiniketan and GBPUA&T, Pantnagar, respectively with an illustrious academic record of excellence. Subsequently, he completed his Ph.D. degree from Indian Agricultural Research Institute, New Delhi in 1999. He started his career as an Agricultural Research Scientist in 2001 and is presently working as Pr. Scientist at ICAR-IISWC, Dehradun. Dr. Mandal has made outstanding contribution in basic, strategic, applied and location specific research in the field of soil and water conservation.

By virtue of hard work and foresight, Dr. Mandal established himself as a renowned soil scientist of country in the field of soil and water conservation, soil quality, and sustainable soil management. He has, for the first time in India, developed a framework to assess soil loss tolerance limits. His cited pioneering research work encompass assessing soil quality and soil loss tolerance limits, water repellence, developing a new concept for soil testing, identifying soil erosion risk areas and classification of land for soil conservation planning and sustainable land use. He is credited with developing a rapid farmer's friendly method to estimate soil health. As a part of a team of experts working on natural resources management for livelihood augmentation for the rural poor, he generated soil health card for 100 farm families for sustainable soil management and demonstrated more than 1000 field trials of farmers field for introducing high yielding varieties and management of soil through low cost technologies. His pioneering research has led to identification of soil erosion risk areas for conservation planning.

Dr. Mandal successfully organized various conference symposia, workshops and group meetings. He has handled more than 10 research projects as PI and Co-PI sponsored by ICAR, MoRD, BARC and MoWR and delivered invited talks at many national and international conferences. He has more than 75 publications in national and international peer reviewed journals, books, bulletins, manuals, book chapters, report and review papers to his credit.

Dr. Gopal Kumar born in Lakhisarai district of Bihar, completed his B.Sc. (Ag.) degree from Rajendra Agricultural University, PUSA, Bihar and Master's and Ph.D. degrees in Agricultural Physics from IARI, New Delhi. He started his career as a Scientist (Soil Physics and SWC) from ICAR-Indian Institute of Soil and Water Conservation and served at Research Centre-Udhagamandalam, Tamil Nadu. He is presently posted as a Senior Scientist at ICAR-Indian Institute of Soil and Water Conservation, Dehradun.

He is a dedicated researcher of multi-disciplinary approach with excellent academic records and recipient of prestigious ICAR -Outstanding Inter-disciplinary team research award, Young Scientist award (IASWC), Prof. P.D. Mistry award (AAM).

His major contribution includes design and development of 'upward flow filter' suitable for plain agricultural field and 'two components filter' for high slope area and channel runoff, design and development of automatic runoff sampler, advance tools based techniques for watershed planning and quick DPR preparation, new methods of remote sensing based watershed impact evaluation, delineation and characterization of ravine in Gujarat, Rajasthan, Uttar Pradesh and Madhya Pradesh and other works in the field of natural resource management . He has authored 50 research papers in peer reviewed journals, 7 authored and edited books, 5 bulletins and others book chapters, 15-popular articles, 5- leaflets and 2- bulletins.

Dr. Rajesh Kaushal, Principal Scientist (Forestry) at the ICAR–Indian Institute of Soil and Water Conservation (IISWC) (Formerly CSWCRTI), Dehradun has 15 years experience of teaching, research and extension in the discipline of Forestry/Agroforestry. He completed B.Sc. (Forestry), M.Sc. and Ph.D. Agroforestry from Dr. Y.S. Parmar University of Horticulture and Forestry, Nauni, Solan (HP). He obtained specialized training on carbon sequestration from The Ohio State University, USA. He is recipient of University gold medal, Himotkarsh gold medal and Santosh Shiksha Puraskar. He has handled six externally funded projects in different capacities and guided three M.Sc. students. He is resource person under South-South knowledge transfer programme of International Network on Bamboos and Rattans (INBAR). He is the Editor of Indian Journal of Agroforestry and Journal of Tree Sciences and life member of many reputed Scientific Societies. He has to his credit more than 60 research papers and 07 books.

Dr. Anup Das, Principal Scientist (Agronomy), ICAR Research Complex for NEH Region, Lembucherra, Tripura was born in Dharmanagar, North Tripura and obtained his Graduation in Agriculture from CAU, Imphal;Master degree in Agronomy from TNAU, Coimbatore and Ph. D. from IARI, New Delhi. During his 15 years research career, Dr. Das has done innovative and pioneering research and extensive extension in the field of conservation agriculture, organic farming, integrated farming system and climate resilient agriculture; developed package of practices for no-till production of pulses and oilseeds in rice and maize fallow, organic production packages of important crops on cropping system basis, developed and disseminated Integrated Organic Farming System model for valley and sloping lands, raised and sunken bed technology for crop intensification in lowlands, site-specific land-use and watershed based farming systems among many other technologies. Dr. Anup Das has 4 years post-graduate teaching experiences in College of Post Graduate Studies, CAU, Umiam and about 10 years experience as Counsellors, IGNOU Study programme on Agriculture at ICAR RC, NEH, Umiam. Dr. Das has guided and co-guided 17 Masters and one Ph.D. student and published 130 research papers, 16 books, 27 technical bulletins, 55 book chapters and many other documents. He is recipient of many National awards and recognitions like NAAS Fellow, NAAS Associate fellow, ICAR-Lal Bahadur Shastri outstanding Young Scientist Award, ICAR-Swami Sahajanand Saraswati outstanding extension scientist award, ICAR-Outstanding inter-disciplinary team research award, Fakhruddin Ali Ahmed award, Dr. HK Jain award for excellence in agricultural research in NE states, Dr. KG Tejwani Award for natural resource Management among many others. He has served as Head, Division of Crop Production, ICAR Research Complex for NEH Region, Umiam, Meghalaya (2016-2018), Secretary of Indian Association of Hill Farming, Umiam (2015-2018), organized more than 100 trainings and events including Summer/Winters schools/Refresher courses, Farmers fair, Field days, National and International Conferences, Workshops, Brainstorming sessions, etc. He has also served as member, National Level monitoring team-NFSM and member Secretary, NMSHE Task Force -6 for north-east India. Presently, Dr. Das is Chief Editor of Innovative Farming and Co-Chief editor of Indian Journal of Hill farming and reviewers of more than 25 National and International journals. He has served as Editor of Indian Journal of Soil conservationists, Indian Journal of Agronomy and Indian Journal of Hill Farming.

Dr. Raman Jeet Singh, presently Scientist (Agronomy) at ICAR-Indian Institute of Soil and Water Conservation (IISWC), Dehradun (Uttarakhand) hails from Sri Ganganagar district of Rajasthan, India. He earned his Bachelor's degree in Agriculture from Rajasthan Agricultural University (RAU), Bikaner, Rajasthan; Master's and Ph.D.degrees in Agronomy from ICAR- Indian Agricultural Research Institute (IARI), New Delhi, India.

Dr. Singh has a brilliant career and experience of more than 10 years in research, training and extension in the field of Conservation Agronomy for sloping crop lands of North-Western Himalaya. Currently, he is studying hydrological behaviour of different land-uses in hilly areas, evaluation of organic and inorganic farming for resource conservation and sustained productivity of different cropping systems, use of agro-geo-textiles in conservation agriculture practices particularly green water technology, validation and refinement of resource conservation technologies in the farmers' fields of North-Western Himalayan region of India. He is also involved in capacity building programmes in the field of soil and water conservation for various categories of stakeholders. Besides this, various international and national peer reviewed journals have identified him as a reviewer for reviewing manuscripts related to resource conservation. He is also serving as the Editor of Indian Journal of Soil Conservation. He is recipient of prestigious Australia Endeavour post doc research fellowship-2016 and did research work on simulation of rainfed conservation agriculture practices at CSIRO- Life and Environmental Sciences, Floreat, Perth, WA. In 2018, he received scholarship from International Soil and Tillage Research Organisation (ISTRO) to attend 21st ISTRO conference in Paris, France. He has published 37 national and international research papers besides 20 popular articles. He has also published two books and four book chapters.

त्रिलोचन महापात्र, पीएच.डी.
एफ एन ए, एफ एन ए एस सी, एफ एन ए ए एस
सचिव एवं महानिदेशक
TRILOCHAN MOHAPATRA, Ph.D.
FNA, FNASc, FNAAS
SECRETARY & DIRECTOR GENERAL

भारत सरकार
कृषि अनुसंधान और शिक्षा विभाग एवं
भारतीय कृषि अनुसंधान परिषद
कृषि एवं किसान कल्याण मंत्रालय, कृषि भवन, नई दिल्ली 110 001
GOVERNMENT OF INDIA
DEPARTMENT OF AGRICULTURAL RESEARCH & EDUCATION
AND
INDIAN COUNCIL OF AGRICULTURAL RESEARCH
MINISTRY OF AGRICULTURE AND FARMERS WELFARE
KRISHI BHAVAN, NEW DELHI 110 001
Tel.: 23382629; 23386711 Fax: 91-11-23384773
E-mail: dg.icar@nic.in

Foreword

The North Eastern region of the country is blessed with bountiful natural resources, forest wealth, agro biodiversity and traditional knowledge system. This region comprising of eight states – Assam, Arunachal Pradesh, Meghalaya, Manipur, Mizoram, Nagaland, Sikkim and Tripura covers a total geographical area of 26.21 mha. It is one of the 12 mega biodiversity hot spots in the country. The potential of these resources can be harnessed best through the use of sustainable soil and water conservation measures to reduce the land degradation problems and improve the productivity of natural resources.

However, this region faced a lot of challenges arising from erratic climatic situation, shrinking common resources, inequalities rising from destruction of biodiversity and human induced interferences with nature. Thus, the best way to have a secure livelihood for the farmers of North-Eastern states is to make balance between the conservation practices and improved agricultural practices. The North-Eastern states through their traditional knowledge based farming system have strived hard to strike the balance. Well established scientific system like *Alnees nepalensis* based farming system, Zabo farming, *panikheti* and poni based farming in plain areas have been well-adopted by farmers. Also, potential of organic farming has been tapped to the fullest with Sikkim becoming the first organic state in India. However, problems like "*Jhum*" cultivation which has devasting consequence on natural resource in still widely practiced in these areas.

Conformingly, the research carried out over the last five decades have clearly established that resource conserving technological interventions, water harvesting and improved cultivation practices have the potential to improve utilization of natural resources and significantly enhance crop productivity in the country. With this backdrop, a collection of field approaches for implementation of soil and water conservation technologies and improved agricultural production based on research information are presented in this book, entitled "**Sustainable Land Management in North Eastern Himalaya**" being published by the Indian Association of Soil and Water Conservationists and ICAR-Indian Institute of Soil and Water Conservation, Dehradun.

With the back drop of climate change and extreme weather events affecting the area which has almost 60% of the area under rainfed agriculture is highly vulnerable to climate change. Thus, amalgamation of various resources conservation practices which has been reflected in this book would be most appropriate for different stake holders and scientific fraternity to rise to the occasion and continue their efforts in the direction for conservation of the natural resources rich biodiversity and heritage of the North-Eastern India.

I congratulate the contributors and editors for having brought out this useful book.

Dated : 28th December, 2018 **(T. MOHAPATRA)**

New Delhi

Preface

Burgeoning population growth and consequent pressure on natural resources and ecosystems warrant effective Natural Resource Management (NRM), soil and water conservation and participatory watershed management. Although the NRM has long history that progressed through various State and Central agencies like ICAR-IISWC (formerly CSWCRTI), NABARD, State Departments, NGOs, *etc.* and significant research and developmental activities have already been undertaken in different regions, various stakeholders of the natural resources and their management lack awareness on the associated issues and technologies, which points out the need of widespread sensitization. The continuing and poorly assessed impacts of human activities on land degradation, availability of water, sustainability of production systems and upstream-downstream linkages are point to this.

Here is a need to assess prevailing resources, opportunities and constraints in given agro-ecosystems by revisiting the past so as to provide a common platform to manage resources optimally, resolve the constraints and renew way forward for ensuring sustainable future jointly with shared responsibilities of all the stakeholders meaningfully. At this context, the Conference "Farmers First for Conserving Soil and Water Resources-2017 (FFCSWR-2017)" brought out wisdom and experience of farmers on the management and conservation of soil and water, forests, streams and rivers complementing the research outputs and scientific knowledge to solve prevailing problems by harnessing potentials and available solutions for the sustainability of the region.

The present book contains selected papers presented in FFCSWR-2017 and is compilation of various topical issues on the prime concerns of natural resource conservation in arable and non-arable lands, soil-water resource development and conservation, diversification of crops and farming systems, management of various ecosystems, people's participation, livelihoods, socio-economic development, success stories, etc. with reference to Assam, Arunachal Pradesh, Nagaland, Manipur, Mizoram, Meghalaya, Sikkim & Tripura which occupies 26.21 mha of land. Papers from other state relevant to resource conservation have also been included in the present compilation having common interests and implications for the regional development are also included. The book is organized in 5 sections: Natural Resource Conservation Technologies, Water Resource Development and Improved Irrigation

Systems, Alternate Land-use Systems, Participatory Watershed Management Programs and Success Stories and Brief Proceedings and Recommendations of the FFCSWR, 2017.

Various papers in this book address not only the issues of combating land degradation but also water management, soil moisture conservation, conservation of other natural resources, improved agriculture production technologies, crop diversification, effective use of degraded lands through suitable alternate land-use systems, improved cropping pattern, modern irrigation systems, soil amendments, *etc.* The book also includes few success stories on resource conservation in different regions. The book suggests improved practices for major crops, such as maize, paddy, wheat, legumes, pulses and mustard, promotion of watershed management and accommodation of ITKs and farmers' wisdom based on experiences of live models and success stories of participatory research and assessment of watershed management programs. Thus, the book has brought out various issues and solutions for intricacies related to participatory planning and execution of appropriate land management practices and improved technologies drawing references from N-W India as well as similar agro-climatic situations.

This book is of much importance particularly at the time when critical introspection on the impacts of soil and water conservation programs and developed technologies vis-à-vis the future needs of resource conservation, farmers and end-users and policy supports are increasingly becoming necessary to reverse factors that cause degradation and production loss. The comprehensive compilation of many ground realities and possible solutions described in this book will immensely benefit farmers, planners, administrators and all other stakeholders of the NRM.

We hope, the readers will find the collection of papers of this book interesting and useful for research and extension endeavours in natural resource conservation and management.

Editors

Acknowledgements

We are thankful to Dr T. Mohapatra, Secretary, Department of Agriculture Research and Education (DARE) and Director General (DG), Indian Council of Agricultural Research (ICAR), New Delhi for his constant inspiration to work with farmers and in the fields that inspired us to organize this conference, Farmers First for Conserving Soil and Water Resources in Northern Region, 2017 (FFCSWR 2017) and bring out this useful book from selected papers. We thank Dr. K. Alagusundaram DDG (NRM) for his consistent support and guidance in conductig the FFCSWR-2017.

Authors are also thankful to Dr. S.K. Chaudhary, ADG for his keen interest and encouragement for bringing out this publication.

The whole hearted participation from the following organization is highly appreciated and their valuable contribution as papers in this compilation is thanked.

The participants from the following Institutes/Organization were also recorded:

- ICAR-Indian Institute of Soil & Water Conservation, Dehradun.
- Assam Agricultural University, Jorhat
- Central Agricultural University, Pasighat, Arunachal Pradesh
- Orissa University of Agriculture & Technology, Bhubaneshwar
- ICAR-Indian Agricultural Research Institute, New Delhi
- ICAR-Indian Institute of Soil Science, Nabi Bagh, Berasia Road, Bhopal, MP
- ICAR Research Complex for NEH Region, Umiam
- National Bureau of Soil Survey and Land-use Planning, Nagpur and regional Stations at Jorhat and Kolkata
- SV College of Agricultural Engg. and Technology, Raipur
- ISRO-Indian Institute of Remote Sensing, Kalidas Road, Dehradun
- Krishi Vigyan Kendra, Longleng, Nagaland
- ICAR-Indian Grassland and Fodder Research Institute, Jhansi

- ICAR-National Organic Farming Research Institute, Tadong, Gangtok, Sikkim
- ICAR-NRC on Mithun, Porba, Phek, Pfutsero, Nagaland
- State Departments of Soil Conservation/Agriculture of North-Eastern Region.
- Soil Conservation Department, Govt. of Assam.

The authors sincerely thank the Ex-Dy. Director General (DDG, NRM), ICAR, New Delhi, Dr A.K. Sikka for his benevolent support and whole hearted guidance throughout the organization of the conference that culminated into this compilation. Dr V.N. Sharda Ex-Member ASRB, New Delhi have been continuing to be instrumental in success of various activities of the Indian Association of Soil & Water Conservationists (IASWC) and in bringing out this publication well with their lasting association with the Association.

The Conference was sponsored by the following organizations besides the organizers, the IASWC and ICAR-IISWC, Dehradun:

- Indian Council of Agri. Research (ICAR), New Delhi.
- Ministry of Agriculture & Farmers Welfare, Govt. of India, New Delhi
- NABARD, Department of Economic Analysis & Research, Mumbai
- National Biodiversity Authority (NBA), Chennai.
- Soil Conservation Department, Govt. of Nagaland.

Innovative farmers who shared their wisdom, lead speakers, special guests, awardees of the IASWC and learned participants are the origin for this publication. All the authors of the papers included in this book are sincerely thanked.

The technical supports of Sh. Amit Singh Rana of IASWC in typesetting this text are generously acknowledged. The assistance rendered by Sh Ravi Mathpal, IASWC and Sh. Ajay Kumar are sincerely acknowledged. The Satish Serial Publishing House, New Delhi who made to publish this book need appreciation.

Editors

Contents

Chapter 1

Watershed Development in Different Agro-ecological Regions: Examples

Prasanta Kumar Mishra

Introduction

Production capacities of our land resource are deteriorating in the pace of rapid population growth, misdirected agricultural practices, and widespread land degradation. Land degradation is mainly attributed to water erosion, resulting into loss of soil fertility, and vegetation and groundwater depletion. Land degradation can be rectified through holistic watershed approach. Improving agricultural production and restoring ecological balance are the twin objectives of watershed programmes. Watershed programmes are recognized as potential engines for agricultural growth and sustainable development. Success and sustainability of these programmes are directly related to collective action for conserving natural resources to enhance crop productivity, and livelihoods for sustainable income generation. The treatment of any watershed is expected to improve agricultural as well as livestock productivity, and livelihood of the beneficiaries.

Keeping in view the problems of different agro-ecological regions, watershed projects were undertaken by ICAR-Indian Institute of Soil & Water Conservation (formerly Central Soil & Water Conservation Research & Training Institute) in different agro-ecological regions of the country during 2008-2014 (Table 1). The project concentrated on capacity building of the beneficiaries in conservation and production aspects with improvement in basic livelihood through different livelihood activities that sustain them, especially during drought years by sustainable use of land and water resources. A blend of the recommended location specific technologies and local level innovations were attempted under the project. This paper presents a comprehensive assessment of the bio-physical and socio-economic impacts of various interventions in the watersheds for augmenting socio-economic/ livelihood security of stakeholders.

Salient Features of Watersheds

These watersheds cover a range of annual rainfall conditions varying from 486 to 835 mm in western and central India, 485-600 mm in semi-arid tracts of southern India, and 1452-1600 mm in eastern India and north-western Himalayan region. The elevation ranges from 72-315 msl in the plain regions to 660-1925 msl in the hilly or sub-mountaineous topographic regions. Average slope varies from 2.2% to 67%. Average landholding size ranges from 0.52 ha to 2.8 ha. A summary of basic features of selected watersheds for implementation is given in Table 2. The present and proposed land-use plan for the selected watersheds is summarized in Table 3.

Table 1: Location and features of study watersheds implemented by IISWC and its Research Centres

Location with State	Watershed name	Village covered	Watershed area (ha)	Treatable area (ha)	Budget Latitude (Lakh)	Longitude
Dehradun (Uttarakhand)	Ashti	Ashti, Dabra, Tangri, Jamuwa	473.0	421.0	47.16	30° 39' 10″ to 30° 40' 35″ N 77° 47' 34″ to 77° 49' 59″ E
Agra (Uttar Pradesh)	Jalalpur	Deori, Jalalpur and Dhanina	697.5	639.5	71.06	26° 49' N to 26° 51' N 77° 32' 30″ to 77° 35' 30″ E
Bellary (Karnataka)	Ramasagara	Ramasagra, Devasamudra, Hanumapur and Venk atapur	485.0	433.4	52.00	14° 49' 31″ to 14° 50' 42″ N 76° 47' 32″ to 76° 49' 16″ E
Chandigarh (UT)	Kajiyana	Kajiyana, Dhattogran, Nala Dakrog & Janouli	509.0	482.2	54.00	30° 46' to 30° 48' N 76° 56' to 76° 59' E
Datia (Madhya Pradesh)	Jigna	Jigna, Imaliya and Tiwaripura	620.0	535.0	62.16	25° 37' 00″ to 25° 39' 30″ N 78° 20' 30″ to 78° 23' 30″ E
Koraput (Odisha)	Lachhaputra-ghati	Lachhamani, Kandaputraghati and Ariputraghati	601.0	557.0	54.12	19° 45' 30″ to 19° 47' 20″ N 82° 56' to 82° 58' E
Kota (Rajasthan)	Dhoti	Dheoti	677.0	660.0	75.00	24° 55' 58″ to 24° 57' 38″ N 76° 31' 22″ to 76° 33' 11″ E
Udhaga-mandalam (Tamil Nadu)	Ayalur	Mallipalayam, Kulaimuppanur, Palapalayam, Semmandam-palaiyam and Pulliyangadu	782.0	708.4	87.48	11° 22' 16″ to 11° 25' 19″ N 77° 22' 43″ to 77° 24' 10″ E
Valsad (Gujarat)	Vejalpur Rampura	Vejalpur, Rampura, Garot, Chikhlod and Ramsoli	775.0	624.0	87.24	22° 58' 58″ to 23° 00' 29″ N 73° 05' 51″ to 73° 08' 19″ E
		Total	**5619.5**	**5060.5**	**590.22**	

Table 2: Summary of basic features of selected watersheds

Particular	Unit	Watershed								
		Ashti	Jalalpur	Ramasagara	Kajiyana	Jigna	Lachaputra ghati	Dhoti	Ayalur	Vejalpur Rampura
Elevation range	m msl	660-1925	184-217	489-693	500-980	240-280	900-1258	900-1258	250-315	72-97
No. of families	No.	80	292	175	151	403	315			
Human population	No.	901	1845	1056	867		992	1612	3610	3484
Av. land holding	ha	1.23	2.4	2.03	0.95		0.52	2.8	1.2	
Cultivable land	ha	153.9	589	355.5	144	365.3	162	561.67	782	597
Livestock population	No.	345 + 890 goat, sheep and poultry	1537	1240	666	3845 + 3066 goat	660	2030	2796+ 12000 poultry	1066
Farm families classification based on land holding size										
Landless	No.	2	30	25	8		84			
Marginal (< 1.0 ha)	No.	46	87	32	103	150	114	11	32.50%	172
Small (1-2 ha)	No.	18	117	79	24	200	91	22	50.20%	133
Medium (2-4 ha)	No.	11	57	49	11	150	22	55	8.80%	175
Large (>4 ha)	No.	3	31	15	5	50	4	105	8.50%	
Summer temp.			48	36	38.5		35.8	49	37.9	40
Winter temp.	°C		1	14	6.4		7.6	6.5	20	12
Rainfall	mm	1600	486	485.8	1220	835.5	1452.2	874	600	812
Monsoonal rainfall	%	75-80	85	72.2	80		81	90	51	94
Rainy days	No.		35	30.4	41	39	77			
Stream length	m	17481	7201	7300	21000		42900		14659	20336
Perimeter	km	10.95	12.5	10.67	11	11.49	10.6		14.58	11.59
Average watershed slope	%	67	2.3	2.2	7.3		12		< 7	< 5
Max. elongated length	m	3740	5100	2750	4700	4800	2915	3125		4239
Max. elongated width	m	2568	2250	1747	1575	1650	2455	2625		2745
First order stream	No.	20	8	5	17		65		12	20
Second order stream	No.	4	3	2	5		14		3	6
Third order stream	No	1	1	2	1		3		1	1
Fourth order stream	No	-	-	-	1	-	2	-	-	-
Mean stream lengths										
First order	m	553	3351	4700	9800		500		394	533
Second order	m	854	850	1950	8000		357.1		2192	659
Particular	**Unit**	**Watershed**								
		Ashti	**Jalalpur**	**Ramasagara**	**Kajiyana**	**Jigna**	**Lachaputra ghati**	**Dhoti**	**Ayalur**	**Vejalpur - Rampura**
Third order	M	3013	3000	650	2325		1316.7		3359	5719
Fourth order	m	-	-	-	875	-	725	-	-	-
Drainage density	km km^{-2}	3.7	1.029	1.52	4.2	2.31	7.14	3.03	1.87	2.62
Compactness ratio		1.42	-	1.36	1.35	1.5	1.22	0.939	1.47	2.32
Elongated ratio		0.81	-	1.57	0.585	0.57	0.922	1.06	0.52	0.74
Circulatory ratio		0.5	-	0.53	0.794	0.56	0.67	0.885	0.46	0.73
Form factor		0.27	-	0.63	0.226	0.26	0.67	0.69	0.21	0.43
Relief ratio		0.34	-	0.015	0.101		95.46	0.003	0.01	0.74
Community land	ha	44.9+ 12.0	40.6	Nil	-		-	63.98		

Table 3: Present and proposed land-use plan for selected watersheds

Land use	Watershed area (ha)															
	Ashti		Jalalpur		Ramasagara		Jigna		Dhoti		Lachha-putraghati		Ayalur		Vejalpur-Rampura	
	Pres	Prop.	Pres.	Prop.	Pres.	Prop.	Pres.	Prop.	Pres.	Prop.	Pres.	Prop.	Pres.	Prop.	Pres.	Prop.
Agriculture	**153.9**	**153.9**	**499.15**	**499.15**	**331.09**	**293.27**	**407.3**	**518.4**	**561.67**	**561.67**	**153.2**	**161.2**	**708.4**	**735.9**	**597.0**	**597.0**
Rainfed	143.7	128.1	380.77	346.77	258.9	175.01	303.5	388.4	421.67	425.67	127.2	109.2	430.5	447.8	545.0	530.0
Crops	143.7	128.1	380.77	338.52	257.9	167.01	271.5	235.8	421.67	421.67	121.2	95.2	430.5	426.6	531.0	517.0
Agro-forestry/ agri.- horticulture / fodder development			Nil	8.25	-	-		40.7		4.0		8.0	-	5.2		13.0
Dryland horticulture/ culturable waste land			-	-	1.0	8.0	-	-	-	-	6.0	6.0	-	16.0	14.0	
Fallow land	-	-	-	-	24.39	61.91	26.1	8.6	103.0	-	68.9	60.9	-	-	-	-
Open srcub	-	-	-	-	-	-	124.8	103.3	-	-	-	-	-	-	-	-
Irrigated	10.2	25.8	118.38	152.38	72.19	118.26	93.8	130.0	37.0	136.0	26.0	52.0	277.9	288.1	52.0	67.0
I Assured	10.2	10.2	39.46	73.46	-	-	-	-	-	-	26.0	52.0	145.1	175.0	50.0	50.0
II Partial	-	15.6	78.92	78.92	72.19	118.26	93.8	130.0	37.0	136.0	-	-	121.8	99.1	2.0	17.0
Coconut garden	-	-	-	-	1.0	2.0	-	-	-	-	-	-	11.0	14.0	-	-
Reserve forest	46.2	46.2	-	-	-	-	9.1	9.1		366.0	366.0					
Civil soyam	210.0	210.0	-	-	-	-	-	-	-	-	-	-	-	-	-	-

Grass land	12.0	12.0	-	-	18.69	18.69	-	-	-	-	8.4	8.4	-	-	-	-
Community forests	44.9	44.9	-	-	-	-										
Waste land/ Barren rocky land			**180.95**	**180.95**	**97.0**	**97.0**	**31.6**	**31.6**	**63.98**	**63.98**			**32.3**	**4.8**	**130.0**	**130.0**
Aforestation			Nil	24.0	-	8.0						6.4		4.8		2.0
Pasture			Nil	116.35	-	28.16					8.2	8.2				15.0
Silvo/silvi - pasture	-	-	-	-	-	-	-	-	-	-	-	1.0	-	-	-	1.0
Horticulture	-	-	-	-	-	-	-	-	-	-	-	-	-	-	-	2.0
Bamboo plantation	-	-		--	-	-	-	-	-	-	-	1.5	-	-	-	7.0
Untreatable			40.6	40.6	-	-										103.0
Fuel fodder	-	-	-	-	-	-	-	-	-	-	-	2.0	-	-	-	-
Biodiesel	-	-	-	-	-	-	-	-	-	-	-	1.5	-	-	-	-
Water body/wetland / *nala* & drain / roads			-	-	7.20	7.20	16.7	16.7	34.35	34.35	75.0	75.0	17.4	17.4	34.0	34.0
Village land/ Others	**6.0**	**6.0**	**17.4**	**17.4**	**2.0**	**2.3**	**44.2**	**44.2**	**17.0**	**17.0**			**23.9**	**23.9**	**14.0**	**14.0**

Watershed Interventions

Watershed treatments were planned keeping in view the specific problems of the watershed and local people's preferences. The technology interventions broadly focused on achieving objectives of (i) reduction in runoff and soil loss, thereby resulting in more moisture conservation and controlling land degradation, (ii) creation of storage for harvested water, (iii) inducing groundwater storage, (iv) increasing crop and water productivity, (v) crop diversification through vegetable crops, horticulture and agro-forestry, (vi) overall environmental sustainability through Drainage Lying Treatment (DLT) and bio-engineering measures, and (vii) livelihood support through income generating activities. Various suitable interventions implemented in these watersheds are presented in Table 4.

Table 4: Region-wise watershed interventions for various land-use categories.

Hilly/Sub-mountaineous region	Plain region
Arable land	
Stone risers construction	Contour bunding / Field bunding
Water harvesting structures	Contour furrows
HDPE pipe line	Irrigation tanks/Farm pond
Inlet chambers	Land levelling
Repair of guhl/tank	Micro-irrigation
Micro-irrigation	
Non-arable lands	
Staggered contour trenching	Staggered contour trenching
Recharge pond *(chaal)*	Water harvesting structure
Road side erosion control/landslide control	Recharge filter
Vegetative filter strips	Recharge pond
Use of geotextile material and plantation	
Drainage line treatment	
Live/brush wood check dams	Live/brush wood check dams
Loose boulders check dams	Loose boulders check dams
Gabion structures/drop spillway	Gabion structures/drop spillway
Retaining wall / retards	Revetment
Production and Micro Enterprises	
Vegetable and horticulture	Vegetable and horticulture
Agriculture crops	Agriculture crops
Forestry/Agro- forestry	Forestry/Agro- forestry
Live stock management	Livestock management
Nursery/polyhouse	Nursery/polyhouse
Integrated farming system (IFS)	Integrated farming system
Organic farming	Organic farming
Livelihood support system	
Tailoring, masonry, carpentry, mechanic, etc.	Tailoring masonry, carpentry, mechanic, etc.
Fish farming, poultry, piggery, apiculture, rice fish culture	Fish farming, poultry, piggery, apiculture, rice fish culture
Animal vaccines, medicines, medication, etc.	Animal vaccines, medicines, medication, etc.

Impact of watershed development work

Data on bio-physical and socio-economic parameters were collected through field visits, detailed resource survey, household survey, PRA techniques, meetings, interviews and FGDs during pre-project and post-project implementation of the watershed project. Periodic monitoring and measurement of hydrological, soils, growth parameters and yield of crops, horticultural and forest plants, land-use, and social and economic parameters were collected. Siltation behind check dams, DLTs and ponds was also measured periodically at selected places during the implementation phase of the watershed. Besides biophysical data, socio-economic data in terms of contribution, change in income, income from SHGs, participation of the community in different activities, etc., were collected through pre-tested questionnaires and interviews. Impact analysis of various activities in achieving the stated objectives is presented as follows :

Objective / Parameter	Impact
Reduction in runoff and soil loss	Reduction in runoff (field scale) – 80% Reduction in soil loss (field scale) – 84%
Creation of storage for harvested water	0.43 – 24.2 ha-m Increase in cropping intensity – 11%
Inducing groundwater storage	Average rise in water table – 0.18-4.0 m (Summer) 12.8-18.3 m (Post monsoon)
Increasing crop and water productivity	Increase in crop productivity- 28-83% (various crops) Increase in water use efficiency- 15-26% (kharif crops) 22-32% (Rabi crops) Increase in soil quality – 6.7-11.5 %
Crop diversification through vegetable crops, horticulture and agro-forestry	Increased area under horticulture/agro-forestry-5-24ha Reduction in fallow area – 13%
Overall environmental sustainability through DLT and bio-engineering measures	Reduction in runoff (watershed scale) – 40% Reduction in soil loss (watershed scale) – 17% Average rate of silt deposition- 13.6 t ha^{-1} yr^{-1} Overall increase in vegetation cover- 4-9%
Livelihood support through income generating activities.	Income earned by SHG's – Rs. 11500-50400 per year Overall increase in income- 5-13%

Lachhaputraghati watershed- A case study

Location

Lachhaputraghati (LPG) watershed is located in Pottangi Tehsil of Koraput district in Odisha state. The watershed is 20 km away from Semiliguda town and 45 km away from Koraput district headquarter. The geographical location is $82^{o}56^{1}$ to $82^{o}58^{1}$ E longitude and $19^{o}45^{1}30^{o}$ to $19^{o}47^{1}30^{o}$ N latitude with an elevation range of 900 m to 1258 m above msl. The total area of the watershed is about 601.24 ha with undulating to steeply sloping (up to 50%) topography. Out of the total geographical area of 601.24 ha, maximum area is under degraded forest (61%) followed by the net cultivated area (20.15%), current fallow (11.5%), area under non-agricultural use (6.0%) and area under pasture land (1.4%).

Basic resources

Climate of the watershed is warm and humid with normal annual rainfall of 1452.2 mm received in 77 rainy days. About 81% of the total rainfall is received during June to September (south-west monsoon). Bright sunshine hours vary from 1.84 to 3.98 and 6.29 to 9.04 during the monsoon and the post monsoon season, respectively. The average evaporation rate is 3.7 mm day^{-1} with maximum in May (6.2 mm day^{-1}) and minimum during the month of August (2.1 mm day^{-1}). Water balance study showed that surplus water is available for agricultural use between the month of May and October with a length of growing period of about 170 days.

The soils are red with sandy clay loam in texture, acidic in reaction, medium in organic carbon (0.69%), soil available nitrogen (288 kg ha^{-1}) and phosphorus content (11.1 kg ha^{-1}) and high in potassium content (313 kg ha^{-1}). The LCC of the watershed revealed that maximum area is under Class III (43.1%) followed by Class VI (22.6%) and Class VII (20%). The Class II and IV account for 6.6% and 7.7% of the total watershed area, respectively. Class II is under paddy cultivation and the majority of the Class III & IV is under rainfed crops of paddy and ragi, whereas Class VI and Class VII land is under degraded forest and shifting cultivation. Out of the total geographical area of 601.24 ha, maximum area is under degraded forest (61%) followed by the net cultivated area (20.15%), current fallow (11.5%), area under non-agricultural use (6.0%) and area under pasture land (1.4%). Watershed has a total population of 992 living in 315 households. Tribal population accounts for about 66% of the total population of the watershed. Major occupation is agriculture providing family income of Rs. 2,500 per month.

Interventions in the watershed

Various interventions were undertaken in the watershed based on problems, needs, priorities of the watershed community, and their technical suitability and economic viability. The watershed development activities taken in the watershed are soil and water conservation measures in arable lands, water resource development, productivity enhancement activities, entry point and income generation activities and community organization including capacity building (Table 5).

Table 5: Natural resource management, production, income generating activities, entry point activities and capacity building activities in Lachhaputraghati watershed

S.No.	Activity	Unit	Quantity
I. Conservation Measures			
1.	Vegetative filter strips	Running meter	300
2.	Field bunding	ha	32.7
3.	Hedge plantation	ha	17
4.	Stone bunding	ha	9
5.	Trenching	ha	13
II. DLT Measures			
1.	Live check dam	No.	35
2.	Brushwood check dam	No.	30
3.	Live brushwood check dam	No.	44
4.	Gabions	No.	13
5.	Stream bank stabilization	Running meter	1124
III. Productivity Enhancement			
1.	Agri-horticulture system	ha	8
2.	Bamboo plantation	ha	1.5
3.	Fuel and fodder plantation	ha	2
4.	Biodiesel plantation	ha	1.5
5.	Silivi-pasture system	ha	1
6.	Agronomic interventions	No.	7
IV. Water Resource Development			
1.	Farm pond	No.	6
2.	*Jhola kundi*	No.	20
3.	Renovation of water harvesting structure	No.	2
4.	Renovation of pipeline system	No.	1
V	**Income generating activities**	Group	15
VI	**Entry point activities**	No.	5
VII	**Capacity building**	No.	49

Impact Assessment

Various Indicators showing impact of watershed development work are assessed before and after the watershed interventions and presented as follows:

Table 6: Productivity impact indicators in the tribal participated LPG watershed

Indicators	Unit	Before (2008)	After (2012-13)	Change
Productivity Indicators				
i. Crop productivity	%		9.14	
ii. Crop diversification index		0.55	0.71	30%
iii. Cultivated land utilization index		0.35	0.4	14.3%
iv. Crop productivity index		0.55	0.61	12%
v. Crop fertilization index		0.21	0.3	43%
vi. Watershed productivity (Ragi Equivalent Yield -REY)	kg ha^{-1}	4962	6126	19%
vii. Av. Survival rate of mango	%	68		

Table 7: Resource use efficiency impact indicators in the tribal participated LPG watershed

Indicators	Before (2008)	After (2012-13)	Change
Resource use efficiency indicators RWUE (kg ha^{-1} mm^{-1})			
Av. Cereals	2.14	2.35	9.9
Av. Pulses	1.47	1.63	11.2
Av.Food crops	1.95	2.15	10.2
Av. Oil seeds	1.04	1.11	6.9
Av. Vegetables	28.92	31.55	9.1
Av. Spices	23.44	25.71	9.7
Average for all crops	11.89	12.99	9.3

Table 8: Environmental impact indicators in the tribal participated LPG watershed

S.No	Indicators	Unit	Before (2008)	After (2012-13)	Change
Environmental impact indicators					
1	**Potential Soil Erosion Rate**				
i	Arable	t ha^{-1} yr^{-1}	17.93	15.61	12.90%
ii	Non- Arable	t ha^{-1} yr^{-1}	37.23	30.38	18.40%
iii	WS Average	t ha^{-1} yr^{-1}	30.24	25.03	17.20%
iv	Soil retention capacity of trenches	t ha^{-1}		13.69	
2	**Estimated Runoff**	%	14.68 to 29.92	7.3 to 15.4	
i	Av.runoff	%	24.4	14.6	40%
3	**Water Resources Development**				
i	Storage capacity created	ha-cm	93.91		
ii	Additional area under irrigation	ha	24.2		
iii	Av.water table depth	m	2.97	2.8	5.90%
iv	Av.depth of water in well	m	0.99	1.17	17.80%
4	Density of trees	trees ha^{-1}	7	14	7
5	Induced watershed eco-index		0.04		4%
6	Carbon Sequestration Potential	Years	10	20	
i	C (t ha^{-1} yr^{-1})		2.12	3.40	
ii	C Credit (Rs. ha^{-1} yr^{-1})		2544	4080	

Table 9: Socio-economic impact indicators in the tribal participated LPG watershed

I.	Socio-economic Impact Indicators				
S.No	Indicators	Unit	Before (2008)	After (2012-13)	Change
1.	**Overall People's Participation Index**	%	56		
2.	**Av. Family Income**	(Rs. yr^{-1})			
i	Large		35700	41000	5300
ii	Medium		21854	31700	9846
iii	Small		13750	18770	5020
3.	**Av. Family Expenditure**	(Rs. yr^{-1})			
i	Large		28500	34000	5500
ii	Medium		18600	27500	8900
iii	Small		13500	18300	4800
4.	**Employment Generation**	Man days	14052		
5.	**IGAs** (Annual income per SHG)	Rs.14, 000/- to Rs.40, 000/- (Rs.900 per head)			
6.	**Amount in WDF Account**	Rs.	121252/-		
7.	**The Economic Viability of the Project**				
I	BCR at 10% DR		1.16		
ii	IRR (%)		19.5		

Technical mandays at different phase of watershed

The technical mandays at watershed work phase is workout to 2.3 and 3.0 manday ha^{-1} (71% of the total mandays) for the total and the treatable area in the watershed, respectively (Figure 6). The technical mandays accounts for only 12 and 17% of the total man-day ha^{-1} during the preparatory and the consolidation phase of the watershed, respectively. Technical mandays slightly higher during the consolidation phase due to completion of pending works coupled with data collection and analysis for impact evaluation.

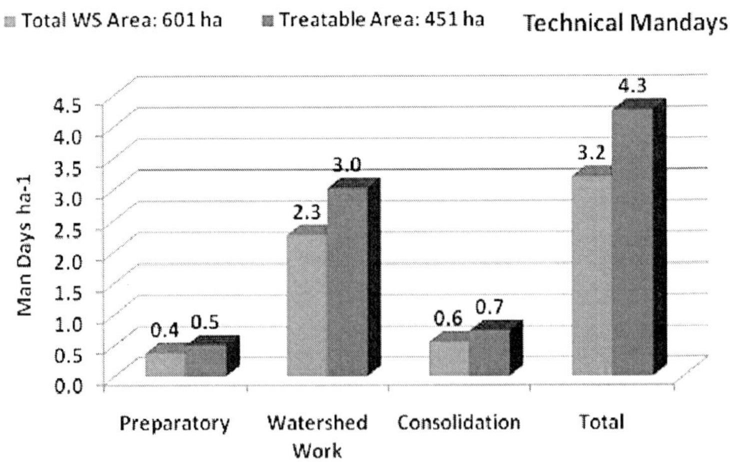

Learning and messages for future of watershed development programme

Based on experiences registered during the project implementation process and impact evaluation analysis, following recommendations are made:

- Need based entry point activities should be taken up as they establish credibility of the Watershed Development Team (WDT) and a rapport with the village community.
- Detailed Project Report (DPR) should be prepared with primary stakeholder active participation, and cover all the proposed works including landless after ground verification.
- Active participation of local community at each and every stage is a pre-requisite for any programme on development and management of watersheds.
- Capacity building of primary stakeholders holds the key to achieve the desired results from watershed development projects.
- Common interest of the community in getting benefits from a particular watershed activity undertaken on community land needs to be exploited for management of common property resources.
- Degraded forest lands on high slopes needs to be protected and developed on priority through community participation and transparent mechanism of sharing of benefits with the community be developed for sustainability.
- Need based conservation measures in arable lands (field bunding) should be adopted, as they play a significant role in in-situ moisture conservation and thereby reducing runoff, soil loss and improving crop yields in rainfed crops, especially during drought years.
- As far as possible ridge to valley sequenced approach should be adopted towards implementation of the watershed works.
- Selection of appropriate site and type of conservation measures and gully control structures to be adopted in non-arable land are essential for reducing runoff peaks, silt deposition, water storage, etc. which helped in gully stabilization and recharges groundwater and borewells in watersheds.

- Management of available water resources through renovation of village pond, construction of water harvesting tanks and improvement in water conveyance efficiency with improved diversified crop production system provides significant impact among farmers of a watershed.
- Productivity enhancement and livelihoods should be the top priority of the watershed development.
- For introducing any alternate land-use system, simple distribution of seeds or tree seedlings leads to failure of the activity. Hence, associated technologies like micro-site improvement, micro irrigation, live fencing, drip irrigation should be given due importance for success of dry land horticulture and agro-forestry system, which plays a major role in improving water productivity.
- Formation of local community based organization with crisp clear roles, responsibilities and benefit sharing mechanism helps to make them sustainable.
- Annual and human development related activities are equally important in watershed development and be given due importance.
- Proper participatory concurrent monitoring and evaluation facilitate mid-term correction in confidence building about demonstrated technology.
- Concept of social fencing needs to be strengthened to regulate grazing habits of animals and develop community land for improving fodder availability to boost milk production.
- Success rate of activities pertaining to water resource creation and their efficient use is very high as farmers show high interest in these activities. These activities should be given high priority and more fund allocation.
- Soil health cards should be prepared under every watershed project and provided to each and every farmer of watershed.
- One of the mandatory conditions for selection of villages for watershed projects should be written agreement of primary stakeholders for contribution towards the Watershed Development Fund (WDF).

- All members of Watershed Committee and staff, such as Watershed Secretary should be given orientation and training to improve their knowledge and upgrade technical/management and community organisational skills to a level that is appropriate for successful discharge of their responsibilities after withdrawal of the watershed development.
- High priority be given to the project activities yielding quick returns to the primary stakeholders individually.
- Livelihood activities introduced in a watershed do not get success, if these are not linked with market intelligence and bank credit. The watershed plus activities must emphasize on such commercial issues after revisiting them.
- Convergence of resources (technical and/or financial) with proper agreed mechanism with state government will facilitate to produce effective results in present and make them sustainable.

❑❑❑

Chapter 2

Promoting Indigenous Knowledge on Agriculture and Allied Sectors in North-Eastern Hill Region, Socio-Economic and Policy Issues

Prof. M. Premjit Singh

Vice-Chancellor, Central Agricultural University, Imphal

The North-Eastern hill region comprises of seven states viz., Arunachal Pradesh, Manipur, Mizoram, Nagaland, Tripura and Sikkim. The region lies between 21.5^0 and 29.5^0 N latitude and 85.5^0 and 97.5^0 E longitude. The total geographical area of the region is 183,813 sq. km (18.38 m ha). This region is endowed with enormous water resources which accounts for nearly 30 per cent of the country's total water wealth. The per capita and per hectare availability of water in this region is highest in the country. But the region experiences severe water shortage during winter and summer months. Majority of the farmers of the hill slopes go for only one crop a year during rainy season and even in the valleys they may not venture to have the crops during *rabi* season (after harvest of paddy), as they do not receive any rains during December and January months and may not have facilities to store water and irrigate the crops.

Though the region receives around 15 per cent of rainfall, the NEH region contributes only 1.5 % to the total food production of the country leading to a perpetual food deficit in the region. This is mostly due to acute shortage of water and very low fertility of the soils of the NEH region.

Soil erosion and nutrient mining are the major causes of land degradation in hilly region. The problem of soil erosion has been aggravated due to a combination of factors like intense rain storms, undulating & steep topography, agricultural intensification and reduction in the land cover due to deforestation. It is estimated that nearly 600 million tons of top fertile soil eroded every year in NEH region and along with its 1.5 million tons of nutrients (NPK) is also lost which is mainly because of steep slopes and also the faulty management of soil and water coupled with high intensity rainfall. Also, there is an imbalance in the quantity of nutrients added to soil by farmers and that removed from soil by crops, due to leaching and other types of losses, thus making the soils impoverished of nutrients.

The predominant practice of Jhuming (Shifting cultivation) on sloppy forest areas has resulted in severe erosion of soil (176.6 Mega grams per hectare per year). Also, traditional practice for cultivation of ginger, turmeric, etc. in raised beds (bun agriculture) in sloppy areas has been estimated to erode 42.3 to 59.5 Mg per ha of top soil. It is estimated that from fallow fields and paddy fields which were earlier under bun cultivation 69.2 to 86.6 Mg per ha per year of soil is lost by erosion.

Indigenous Knowledge

Indigenous knowledge (IK) in the general sense is the knowledge used by local people to make living in a particular environment. Indigenous knowledge can be conceptualized as the knowledge that a local community accumulates over generations of living in a particular environment. It is defined as "a body of knowledge built up by a group of people through generations living in a close contact with nature". Unfortunately many scientists regarded the indigenous knowledge and technologies as irrational and invalid. But the ITKs were found to be practical and utilitarian, and are the result of a more general intellectual process of creating order out of disorder and not simply as a response to particular human needs. The indigenous knowledge has been found to be sustainable and eco-friendly.

In recent years various governmental and non-governmental organizations are attempting to document and validate the knowledge and skill possessed by indigenous communities and also the indigenous technologies developed by them.

Incorporating indigenous knowledge into research and extension activities is the need of the hour, as it can contribute to local empowerment or capacity building and also in attaining self-sufficiency in food production. In addition, it enhances the cultural pride and help to motivate the local people to solve the local problems with local resources.

Though a vast majority of our farmers are toiling under these difficult farming situations over centuries in NEH region, some potential eco-friendly indigenous farming systems have been developed by some innovative farmers in some isolated pockets using their wisdom, ingenuity and skill. These techniques are being practiced over centuries and have been found to be sustainable. These innovative indigenous technologies were developed by making use of the locally available resources. Now there is a need for an in-depth scientific study to understand the science behind them and the secrets of their success.

Some of the indigenous knowledge and technologies on agriculture and allied sectors are being listed and discussed in brief.

I. Agriculture Sector

Some of the indigenous technologies in the field of soil and water conservation being practiced by the farmers of NEH region have been compiled.

Indigenous farming systems of NEH region

1. **Shifting cultivation:** It is still prevalent in NEH region on an estimated area of 1.47 m ha with jhum cycle of 5-10 years. The productivity of the Jhum land is very low (0.87 t/ha). Though shifting cultivation leads to reduction in forest covers and erosion of the soil, it is inevitable to the farmers of the region.
2. **Bun cultivation**: This system of cultivation is mainly practiced in Meghalaya state for growing crops like like ginger, turmeric, potato, sweet potato, etc., In this system, crops are grown on a series of raised beds locally known as "bun" formed along the slope of the hills. The phytomass is placed over raised beds and burnt, thus exposing the land during dry season. The Bun system is more hazardous than slash and burn system since the soil erosion is more severe in these plots.

A) Other indigenous practices

1. Apatani water management system in Arunachal Pradesh

It is a multipurpose water management system developed by the *Apatani* tribe to cultivate paddy and fish together and is practiced over 30 sq km area at an altitude of 1525 m above MSL in lower *Subansiri*

District of Arunachal Pradesh. This system helps in preventing soil erosion, conserves water which can be used for irrigation of paddy land and for fish culture.

2. Yetbung Lingang, Abu Lingang and Linkumare traditional water conservation structures practiced by Adi tribes of Arunachal Pradesh

From the study by scientists of college of Horticulture and Forestry Pasighat under CAU, Imphal found that the Adi tribe of Arunachal Pradesh has very rich traditional knowledge in conserving natural resources. The *Yetbung Lingang, Abu Lingang and Linkumare are the* traditional water conservation structure made of locally available stones/boulders bamboo, takopata and can be used to divert the perennial stream of the area for *panikheti* and winter crop cultivation. They protect the gully from erosion by constructing traditional *Rango* structure. It is like brush wood dam. To support the dam, a structure called *Korang* which is filled with stones is kept in front of the dam. The *tokopata* is used to prevent the passage of sediment.

3. Bamboo drip irrigation system of Meghalaya

The indigenous technique of drip irrigation of betal vines using bamboo was developed by tribal farmers of Muktapur, Jaintia hill district of Meghalaya. In this system, natural stream of water at higher elevation is conveyed through bamboo channels with wooden or bamboo support to the site of plantation through gravity flow. Water distribution to the entire garden is done with the use of bamboo channels, bamboo supports, water diversion pipes and strips. The whole system would help to distribute 15-25 litres of water without any leakage at joints.

4. Roof top water harvesting in Mizoram

The roof water harvesting technique was developed by people living in steep slopy areas of Mizoram long back since people of the state were facing acute shortage of water for drinking and other purposes during summer months. In houses built with sloping roofs, water harvesting structures are made of galvanized iron sheets for collection of rain water and its storage. The harvested water was being stored in tanks made of GCI sheets or galvanized plain sheets, in their court yard in recent years cement tanks are also used.

5. Bench terracing with continuous flow of irrigation for soil and water conservation in Nagaland

Farmers in some parts of Kohima district have been cultivating crops in bench terraces from time immemorial in slopy lands upto 100 % slope and even in rocky lands with just 10-15 cm depth of soil. In this system, the water streams from upper region of the hills are tapped and channelized to irrigate a series of terraces so that water flows continuously from upper terraces to lower ones without soil erosion and maintaining 5 to 8 cm of water in each terrace. Terrace risers (bunds) are usually maintained with stones and the land is prepared by digging with spades and then transplanting of paddy seedlings is taken up.

6. Agriculture with Alders (*Alnus nepalensis*) tree for preventing soil erosion and enriching soil

The farmers in Angami, Chakhesang, Chang, Yimchunger and Konyak areas in Nagaland at varying altitudes cultivate alder tree and use in agriculture.

The seedlings are planted in sloppy land intended for cultivation of crops. The alder grows very fast and at 6-10 years age, the trees are pollarded, the leaves and twigs are burnt and ash is mixed with soil and crops are grown. Subsequently pollarding is done once every 4-6 years. The bigger branches are used as firewood. The trees have extensive root system with nodules which fix atmospheric nitrogen and also prevent soil erosion.

7. Zabo system of farming

The word *Zabo* means impounding water. *Zabo* is an indigenous farming system of Nagaland, originated in Kikuma village, Phek district; practiced in an area of more than 900 ha. This farming system has a combination of forest, agriculture and animal husbandry. In this system, the protected forest land is at the top of the hill, water harvesting tanks in the middle and paddy field at the lower side. Special techniques, such as incorporation of paddy husk on shoulders bunds and puddling are done to control seepage losses of water in the paddy fields. The forest land (1.5 ha or more) is used as catchment area to serve as source of water for the tank.

The forest catchment is not disturbed by cutting and burning of trees but sometimes cattle, pigs and poultry birds are allowed to enter the forest area. In the mid-hills silt retention and water harvesting tanks with earthen embankment (using dug up soil) of 24 m x 10 m x 2 m are made. About 0.2 ha area is used for water harvesting system. The bottom of the tank is properly rammed and the sidewalls are plastered with paddy husk to minimize loss of the through seepage. The open cattle enclosure is fenced with wood and branches of bamboo and farmers keep their cattle inside these enclosures.

The water from the ponds is made to pass through cattle yard before entering rice field. This water carries with it the dung and urine and thus, enriches the paddy field. A small pit is dug in the middle of paddy field and fish fingerlings are put in the field after transplanting. Before harvest the water is drained out, the fish will remain in the pit which is harvested later. The farmers get around 3-4 tons of paddy and 50-60 kg of fish per hectare.

8. Zero tillage cultivation of rapeseed-mustard in rice fallow

In Manipur, rabi crop cultivation is totally rainfed and dependent on the conserved moisture in the soil, as there is no facility for irrigation Zero tillage is the cultivation of crops without giving any kind of disturbance to the soil. It is the practice of not adopting any tillage operation or minimum tillage operation prior to sowing with the assumption that moisture losses increases once the soil is open and seeds may not be able to germinate.

The zero tillage cultivation of rapeseed-mustard, already being a widely adopted practice among the farmers in Manipur was given emphasis with little modifications and scientific approach as detailed below by the Central Agricultural University, Imphal:

a) Bee pollination (at least 4 bee colonies per hectare)

b) Reduction in seed rate (12 kg per ha from 20-25 kg/ha) used traditionally by the farmers.

c) Use of reduced dose of chemical fertilizers @40:20:20 kg NPK/ha

d) Use of liquid bio-fertilizers, bio-pesticides, plant products, etc.

e) Introduction and use of Zero-Till Seed Drill machine for efficient sowing.

9. ITKs in Animal Husbandry

a) Yak and cattle cross breeding to enhance milk production.

The crossbreeding between yak and local hill cow is being practiced in Arunachal Pradesh. The F1 generation male called as ***DZO*** is found to be sterile, while the female called as ***DZOMO*** is fertile and produces double the quantity of milk than yak. However, the F2 generation *Dzomo* is a poor milk yielder. The local people go upto F1 generation only for enhancing milk production.

b) Use of rice bran mixed with *choak* (rice beer) waste as a pig feed.

c) Production of country cheese (*churpi*) in Arunachal Pradesh – The milk is heated upto 40° C and then cream is separated out by churning. The milk is converted to curd which is churned to separate the butter. A starter is added to skim milk and after incubation the soft *churpi* is separated out.

d) Enhancing the quality (taste) and shelf life of pork by subjecting smoke treatment.

e) The other important practices are:

   i) Use of neem leaves and tobacco powder to control ticks, lice and fly menace in animals

   ii) Use of turmeric rhizome paste to cure skin diseases caused by ecto-parasites in animals

   iii) Use of bark, flower and fruits of silk cotton tree (*semal*) for treatment of skin diseases and ecto-parasites in animals.

10. Indigenous knowledge in Fisheries sector

The farmers of NEH region always adopted the integrated farming system, where the components of animal farming and fishery farming have been practiced together.

Some of the practices developed and adopted over generations are being listed below:

a) Fish culture in low land paddy fields

b) Integrated fish-cum-duck farming

c) Bottle gourd cultivation over fish ponds

d) Use of jack fruit seeds as fish feed

e) Use of *kack-dumur Gobla* in Hindi *Ficus carica* as feed for fish and prawn

f) Control of fish lice using palm tree leaves

g) Use of refugee lata (*Mikania cordata*) extract mixed with green turmeric juice to control Epizootic Ulcerative Syndrome (EUS) disease in fish pond

h) Use of banana leaf ash as substitute to lime for curing EUS disease

i) Enhancing shelf life of fish by smoke treatment.

A vast number of indigenous technologies are still being practiced by farmers in NEH region in isolated pockets which definitely have the socio-economic implications. There is a need to carry out in-depth research the technicalities of these technologies and ways and means of improving them.

The scientists will have to suggest the policy-makers for conserving soil and water resources of the region. In addition the following policy issues and research initiatives have to be taken up in the NEH region for enhancing livelihood security of the farmers in NEH region.

Socio-economic and policy issues

- Promotion of zero tillage technology - since the region has vast stretches of paddy field, promotion of zero tillage and sowing of mustard or any short duration legume during rabi season would definitely increase the income of the farmers. The CAU has done demonstration in farmers fields over an area of 1000 ha on cultivation of mustard in rabi season after harvest of paddy following zero tillage for 5(five) consecutive years.

- Promotion and development of genotypes for the climate change scenario - since the climatic condition of NEH region are changing in recent years, there is need to develop crop vaieties to such changed climatic situation.

- Promotion of indigenous crops which are well-adapted to the different altitudes and soil conditions of NEH region. This will bring diversity in crops and help in reducing genetic vulnerability.

- Improvement of traditional implements to help in mechanization rather than depending on improved machinery.

- Orientation and development of agricultural implements with women centric approach, as women form important segments of agricultural labour.
- Promotion of ITK based disease and pest control as the farmers utilize locally available resources.
- Patenting and registration of GI for such implements and practices of NEH region so that their adoption is taken as pride, thus helping in sustained agricultural production.
- Research on organic agriculture based indigenous materials should be intensified.

❑❑❑

Chapter 3

Land and Water Management for Livelihood Improvement of Small and Marginal Farmers in North-Eastern Hill Region

S.V. Ngachan and B.K. Sethy

ICAR Research Complex for NEH Region, Umiam-793103, Meghalaya

Introduction

The north-eastern region comprising 8 states of the country, namely Arunachal Pradesh, Assam, Manipur, Meghalaya, Mizoram, Nagaland, Sikkim and Tripura, respectively is a physically difficult, topographically differentiated, culturally diversified, socially isolated and economically backward region. The entire region is unique in its physiographic characters of land, varying agro climatic conditions, land tenure systems and cultivation practices different from the rest of the country. The region extends over about 26.2 m ha accounting for 7.9 per cent of country's land area with 45.6 million population of which 21.39 % is tribal (Table 1). About 35 % area in the region is plain excepting Assam, where plains account for 84.44 % of its geographical area. The total forest cover of the region is 49.65% of its geographical area as against the national average of 19.39 %, which is about 25% of the total forest cover in the country. The plain land consists of Brahmaputra and Barak valley in Assam, Imphal valley in Manipur and some flat land scattered between the hills in Tripura.

Table 1: State wise Total Scheduled Caste and Scheduled Tribe (Male, Female, Rural and Urban) Population

States	Schedule Cast Population			Schedule Tribe Population		
	Total	Rural	Urban	Total	Rural	Urban
Arunachal Pradesh	-	-	-	951821	789846	161975
Assam	2231321	1825761	405560	3884371	3665405	218966
Manipur	97042	47563	49479	902740	791126	111614
Meghalaya	17355	11573	5782	2555861	2136891	418970
Mizoram	1218	298	920	1036115	507467	528648
Nagaland	-	-	-	1710973	1306838	404135
Sikkim	28275	20335	7940	206360	167146	39214
Tripura	654918	437993	216925	1166,813	1117566	49247
Total NER	**30,30,129**	**23,43,523**	**6,86,606**	**97,52,260**	**83,85,601**	**13,66,659**

Source: Census, 2011.co.in

Majority of the population of the region is dependent on agriculture and land-based activities. Share of marginal and small farmers is 78.92 per cent. About 58% of the workforce of the region is engaged in agriculture with 252 lakhs operational holdings (Table 2). The average size of land holdings in NEH regions is 1.95 ha, which is higher than the national average. It indicates ample scope for agricultural development in the region which is waiting for second green revolution. The contribution of agriculture to NDP of NEH is 35 % as compared to national average of 34 per cent. The agricultural production system in the region is mostly rain-fed and mono-cropped at subsistence level. Farming is predominantly rice based with exception in the state of Sikkim, where maize is the predominant cereal crop. The net cultivated area varies widely from 3.37% (Arunachal Pradesh) to 26.41% (Tripura) and is approximately 6.8% of the total reported area of the states. Among all the states, maximum culturable wasteland (4.5 M ha) is in Meghalaya which is under the age-old practice of shifting cultivation. The total area & production and average productivity of the region in 2013-14 were 3983.5 '000 ha, 7818.5 tonnes and 1952.8 kg/ha, respectively (Table 3). The productivity of most of the crops is much lower than the national average.

Table 2: State-wise Number of Total Worker, Main and Marginal Worker

State	Total workers			Main worker			Marginal worker		
	Person	Male	Female	Person	Male	Female	Person	Male	Female
Arunachal Pradesh	587657	350273	237384	478721	301109	177612	108936	49164	59772
Assam	11969690	8541560	3428130	8687123	7034642	1652481	3282567	1506918	1775649
Nagaland	974122	547357	426765	741179	442204	298975	232943	105153	127790
Manipur	1159053	665463	493590	855012	554518	300494	304041	110945	193096
Mizoram	486705	290740	195965	415030	263305	151725	71675	27435	44240
Meghalaya	1185619	703709	481910	921575	585520	336055	264044	118189	145855
Tripura	1469521	1045326	424195	1077019	887881	189138	392502	157445	235057
Sikkim	308138	194358	113780	230397	160513	69884	77741	33845	43896
Total NER	18140505	12338786	5374954	13406056	10229692	3176364	4734449	2109094	2625355

Source: Census of India, 2011

Table 3: State-wise Area, Production and Yield of Total Foodgrains

States	Area(*'000 Ha*)		Production(*'000 Tonne*)		Yield (*Kg./Ha*)	
	2012-2013	2013-2014	2012-2013	2013-2014	2012-2013	2013-2014
Arunachal Pradesh	203.8	214.4	364.0	384.6	1786	1794
Assam	2691.5	2660.0	5280.6	5096.8	1962	1916
Manipur	174.8	281.1	336.7	490.6	1926	1745
Meghalaya	132.7	134.1	265.0	320.0	1997	2387
Mizoram	23.8	48.4	41.8	72.8	1756	1506
Nagaland	285.7	309.5	579.1	624.6	2027	2018
Sikkim	68.9	64.9	106.0	102.4	1538	1577
Tripura	267.5	271.1	725.2	726.7	2711	2680
Total NEH	**3848.7**	**3983.5**	**7698.4**	**7818.5**	**1962.8**	**1952.8**
All India	**120778.7**	**125042.0**	**257124.7**	**265043.2**	**2129**	**2120**

Source: Ministry of Agriculture, Govt. of India

Livestock is an integral part of farming, as it fulfills the animal protein requirement of the population as majority of the population is non-vegetarian. The region has over 265.99 and 435.35 lakh livestock and poultry populations constituting 5.19 % and 5.97 %, respectively of total livestock and poultry populations of the country (Table 4). The livestock population in the region is very poor except in the state of Assam, but has wide diversity in the livestock population with rearing of yak, mithun, rabbit, etc. as it is part of tribal lives, rituals and customs. Therefore, the region has enormous potential to enrich commercialization through scientific techniques, improved breeds and management of pastures and common property resources for quality fodder production.

Table 4: Total Number of Livestock and Poultry

States	Rural		Urban		Total (Rural+Urban)	
	Livestock	Poultry	Livestock	Poultry	Livestock	Poultry
Arunachal Pradesh	1366179	2188749	46487	55482	1412666	2244231
Assam	18846897	26837618	235274	378551	19082171	27216169
Manipur	610659	1920770	85113	578746	695772	2499516
Meghalaya	1951969	3379945	5658	20087	1957627	3400032
Mizoram	200901	850755	110955	420598	311856	1271353
Nagaland	806071	1838703	105091	339767	911162	2178470
Sikkim	288979	445734	2647	6232	291626	451966
Tripura	1888782	4092513	47397	180220	1936179	4272733
NER	**25960437**	**41554787**	**638622**	**1979683**	**26599059**	**43534470**
All India	**491685846**	**697895256**	**20371455**	**31314064**	**512057301**	**729209320**

Source: 19th Livestock Census, 2012

The region is endowed with bounty of water resources sharing about 46 % of the total water resources of the country. Though the region receives an annual rainfall of 2000 mm accounting for around 10 % of the country's total precipitation, yet it can be characterized as plenty and scanty in terms of rainfall, which is a major source of water supply. The tentative assessment of this dynamic resource is about 63.6 m ha-m. Out of which 25.44 m ha-m is lost as evaporation, 12.72 m ha-m goes into the soil system including groundwater recharge and storage in soil mass, and 25.44 m ha-m flows as surface, sub-surface and interflow. However, only 0.88 m ha-m is utilized and the rest of the water resources are lost as runoff to the downstreams. In the absence of the appropriate infrastructure, hardly 18 % area is under irrigation and there is absolute water scarcity during November to April both in hills and plains. Hence, the region's water scenario is classified under "economic drought" category. Emphasis should be given on rainwater harvesting, integrated watershed approach for creating supplementary irrigation facilities. In the absence of scientific management of this vital resources from rain and consequent surface flow and underground storages, the water resources goes waste and creates havoc downstream. Much of the enormous water resources remain unutilized due to absence of proper water resource planning and scientific management. The efficient utilization and management of available rainwater is the core issue.

Slash and burn agriculture (nearly 0.88 m ha area) is still practiced in almost all the states. This practice of slash-and-burn agriculture on steep slopes and expansion of agriculture to erosion-prone land results in as high as 76.6 Mg ha^{-1} yr^{-1} of soil loss. Consequently, 80% of the cultivated area is under threat of moderate to severe erosion threatening ecological balance and food security for future generations. The multi-vegetated agricultural hilly watershed converts about 14% of the rainfall into runoff with average sediment concentration of 6 kg m^{-3} in runoff. The average sediment yield from the multi-vegetated hilly watershed is 22 Mg ha^{-1} yr^{-1}, which exceeds the maximum permissible limit of 16 Mg ha^{-1} yr^{-1}. Valuable nutrients are also transported out of watershed with runoff and sediment with an average rate of 33 kg ha^{-1} nitrogen, 0.30 kg ha^{-1} phosphorus, 27 kg ha^{-1} potassium and 750 kg ha^{-1} organic carbon.

The region, once richly endowed with genetic diversity of plants, has been denuded due to human interferences and adoption of unscientific land-use systems. With rapid increase in human and livestock population and the rising demand of food, feed, fuel, fodder, fiber, timber and the other developmental activities, the farmers have been forced to exploit forest land and water resources at sub-optimal level in complete defiance of the inherent potential. This has resulted in progressive decrease of forest cover, loss of biodiversity, serious soil erosion leading to depletion of plant nutrients, degradation and decline in land productivity and its carrying capacity, silting of major river basins causing frequent ûoods in the plains, and drying up of perennial steams as well as ecological imbalances. Gradual degradation of these resources is of primary concern and calls for location-specific measure to efficiently conserve, utilize and manage these resources for optimizing production on sustained basis without adversely affecting its quality.

Despite agriculture is the main economic activity in the region, major impact of green revolution in the irrigated areas of the country; modernization of agriculture has escaped this region, as evidenced by poor adoption of modern technologies, low consumption of fertilizers (as low as 2.05 kg/ha in Arunachal Pradesh to 72.76 kg/ha in Tripura) and other indicators of growth. One of the most important reasons for this slow growth is the lack of location-specific and system-based technologies. Public investments in agriculture in this region for this period are grossly inadequate to be compatible with the requirements of input intensive agriculture. Thus, there is an imperative need to improve the food grain production from this resource-rich region through adoption of multi-pronged strategies to efficiently conserve and utilize the natural resources and ensure livelihood security to the rural and tribal community of the region. There are deficit in cereals, pulses, oilseeds and food grains, thus majority of the population suffers from under-nourishment and depends on other states for their requirements. There is a deficit of 7.3%, 82%, 82 % and 15 % for cereals, pulses, oilseeds and total food grain production, respectively in NEH region. States like Arunachal Pradesh, Nagaland, Sikkim and Tripura in NEH region are surplus in cereals. Pulses and oilseeds are deficit in all the states of the region. To achieve the self-sufficiency in production of cereals, pulses and oilseeds, an additional area has to be brought under cultivation or by increasing the productivity levels of crops significantly through modern production technologies coupled with increasing area under inter-crops, irrigation and by efficient management of inputs and natural resources.

Important Programs for Livelihood Security

National Policy for Farmers

Government has already implemented the National Policy for Farmers (NPF), 2007, which aims to improve economic viability of farming and increase net income of farmers. Policy provisions in NPF-2007 include, inter-alia, asset reforms in respect of land, water, livestock, fisheries and bio-resources; supply of good quality seeds and disease-free planting material, issue of soil health passbooks to the farmers and integrated pest management system; region and crop specific implements and machinery; support services for women; timely, adequate and easy reach of institutional credit at reasonable interest rates and farmer-friendly insurance instruments; support services and inputs like application of frontier technologies; agricultural bio-security system; use of Information and Communication Technology (ICT) and setting up of farm schools to revitalize agricultural extension; coverage of farmers under a comprehensive national social security scheme; effective implementation of Minimum Support Price (MSP) across the country and establishing community food grain banks; development of agricultural market infrastructure and terminal markets for agriculture; curriculum reforms in agricultural universities; special categories of farming like organic farming and contract farming; rural non-farm employment initiative for farm households; and integrated approach for rural energy, etc.

Deen Dayal Antyodaya Yojana - NRLM

Aajeevika-National Rural Livelihoods Mission (NRLM) was launched by the Ministry of Rural Development (MoRD), Government of India in June 2011. Aided in part through investment support by the World Bank, the Mission aims at creating efficient and effective institutional platforms of the rural poor, enabling them to increase household income through sustainable livelihood enhancements and improved access to financial services. NRLM set out with an agenda to cover 7 crore rural poor households, across 600 districts, 6000 blocks, 2.5 lakh Gram Panchayats and 6 lakh villages in the country through self-managed Self Help Groups (SHGs) and federated institutions and support them for livelihoods collectives in a period of 8-10 years. In addition, the poor would be facilitated to achieve increased access to rights, entitlements and public services, diversified risk and better social indicators of empowerment. DAY-NRLM believes in harnessing the innate capabilities of the poor and complements them with capacities (information, knowledge, skills, tools, finance and collectivization) to participate in the growing economy of the country. In November 2015, the program was renamed Deen Dayal Antyodaya Yojana (DAY-NRLM). The overall mission of NRLM can be stated as "To reduce poverty by enabling the poor households to access gainful self-employment and skilled wage employment opportunities, resulting in appreciable improvement in their livelihoods on a sustainable basis, through building strong grassroots institutions of the poor." Under this scheme, an expenditure of Rs.15, 000 per person is allowed on training of urban poor but in North-east and J&K allowable expenditure is Rs.18, 000 per person.

Mahatma Gandhi National Rural Employment Guarantee Act (MGNREGA)-2005

National Rural Employment Guarantee Act, 2005 (or, NREGA) was later renamed as the "Mahatma Gandhi National Rural Employment Guarantee Act" (or, MGNREGA), is an Indian labour law and social security measure that aims to guarantee the 'right to work'. It aims to enhance livelihood security in rural areas by providing at least 100 days of wage employment in a financial year to every household whose adult members volunteer to do unskilled manual work. Another aim of MGNREGA is to create durable assets (such as roads, canals, ponds, wells). Employment is to be provided within 5 km of an applicant's residence, and minimum wages are to be paid. If work is not provided within 15 days of applying, applicants are entitled to an unemployment allowance. Thus, employment under MGNREGA is a legal entitlement. Labour intensive tasks like creating infrastructure for water harvesting, drought relief and flood control are preferred. Apart from providing economic security and creating rural assets, NREGA can help in protecting the environment, empowering rural women, reducing rural-urban migration and fostering social equity, among others. Starting from 200 districts on 2nd February, 2006, the NREGA covered all the districts of India from 1st April, 2008. MGNREGA has great impact on North-eastern states. Tripura retained its top position for the seventh consecutive year in providing jobs under the MGNREGA-a staggering 94.46 days per household in fiscal 2015-16 against the national average of 48.51 days. According to a performance report of the union rural development ministry, Mizoram and Sikkim were second and third by providing 68.55 and 66.97 days of jobs in the fiscal. In sharp contrast, two other north-eastern states Arunachal Pradesh and Manipur were at the very bottom among India's 29 states by providing only 27.47 and 15.65 days of jobs. (MGNREGA scheme: 3 N-E states on top, two at bottom: The financial express: April 27, 2016).

Drought Prone Areas Programme (DPAP)

The basic objective of the programme is to minimize the adverse effects of drought on production of crops and livestock and productivity of land, water and human resources ultimately leading to drought proofing of the affected areas. The programme also aims to promote overall economic development and improving the socio-economic conditions of the resource poor and disadvantaged sections inhabiting the programme areas. The Union Government sanction block-wise new projects every year to programme districts taking in to consideration primarily the DPAP coverage, performance of the on-going projects, capacity to absorb new projects and annual budget outlay, etc. During the year 2005-06, 3000 new watershed projects was sanctioned under DPAP to treat an area of 15 lakh hectares at a total cost of Rs. 900.00 crore over a period of five years.

Integrated Wasteland Development Programme (IWDP)

The Watershed approach has conventionally aimed at treating degraded lands with the help of low cost and locally accessed technologies, such as in-situ soil and moisture conservation measures, afforestation, etc. and through a participatory approach that seeks to secure close involvement of the user-communities. The broad objective was the promotion of the overall economic development and improvement of the socio-economic conditions of the resource poor sections of people inhabiting the programme areas. Many projects designed within this approach were, at different points of time, taken up by the Government of India. The Drought Prone Areas Programme (DPAP) and the Desert Development Programme (DDP) were brought into the watershed mode in 1987. The Integrated Wasteland Development Programme (IWDP) launched in 1989 under the aegis of the National Wasteland Development Board, also aimed at the development of wastelands on watershed basis. All these three programmes were brought under the Guidelines for Watershed Development with effect from 1 April, 1995. Other major programmes were implemented through this approach are the National Watershed Development Project in Rainfed Areas (NWDPRA) and the Watershed Development in Shifting Cultivation Areas (WDSCA) of the Ministry of Agriculture (MoA). The focus of these programmes has, with the advent of the Department of Land Resources (DoLR) shifted to the enhancement of the viability and quality of rural livelihood support systems. While the programmes of DoLR were designed to address areas characterized by a relatively difficult terrain and preponderance of community resources, those of Ministry of Agriculture are expected to aim at increasing production and enhancing productivity in cultivated areas largely privately owned. While the focus of these programmes may have differed, the common theme that underpinned their structure has been the basic objective of land and water resource management for sustainable development of natural resources and community empowerment.

Apart from the above mentioned programmes/schemes, there are many more programmes/schemes being undertaken, which includes Prime Minister's Rural Development Fellows Scheme (PMRDFS), National Rurban Mission (NRuM), Investment Promotional Scheme (IPS), Deen Dayal Upadhyaya Grameen Kaushalya Yojana (DDU-GKY), Prime Minister's Fasal Beema Yojana, Attracting Rural Youth in Agriculture (ARYA).

Resource Conservation Programmes

The Government of India has launched various programmes/ schemes, to increase the productivity and livelihood security in the region through rehabilitation of degraded lands following the approach of watershed management. The prominent national programmes implemented in NEH region are National Watershed Development Programme for Rainfed Areas (NWDPRA), Integrated Watershed Development Programme (IWDP), Integrated Watershed Management Programme (IWMP), River Valley Project and Flood Prone Rivers (RVP/EPRs), Watershed Development Project for Shifting Cultivation Areas (WDPSCA) and National Aftorestation Programme (NAP). Besides these national programmes/schemes, many state Govt. funded programmes were also implemented in some states in NEH region.

In Meghalaya, for instance, the various soil conservation programmes/schemes implemented by the Department of Soil & Water Conservation include both Centrally Sponsored Schemes as well as State Plan Schemes. They are: **i)** Soil & Water Conservation in General Areas and Watershed Management Programme under **State Plan Schemes; ii)** Integrated Wasteland Development Programme (IWDP) under c**entrally Sponsored Schemes; iii)** Watershed Development Project in Shifting Cultivation Areas (WDPSCA) and Accelerated Irrigation Benefits Programme (AIBP) under **additional central assistance; iv)** Rural Infrastructure Development Fund (RIDF) under **NABARD Loan v)** Soil Conservation for enhancing the productivity of degraded lands in the catchment of River Kopili in Jaintia Hills District under Macro-Management Mode of Agriculture Department, Meghalaya and Rashtriya Krishi Vigyan Yojana (RKVY) under **other Government of India Schemes; and vi)** Cherrapunjee Ecological Project- Restoration of Degraded Lands under Sohra Plateau under Special Plan Assistance.

Watershed management techniques has been proved successful in conserving soil resources throughout the country implemented by both, Central and state Government departments in participatory mode. Hence, it is call of the time to implement the participatory Watershed Management programme for management of natural resources.

Tribal Sub Plan (TSP)

The Tribal Sub-Plan was initiated during Fifth Five Year Plan for socio-economic amelioration of the tribal communities. The ST development effort was revamped and invigorated in order to fulfill the constitutional mandates of ensuring better quality of life of the schedule tribes. The basic objective of Tribal Sub-Plan is to channelize the flow of outlays and benefits from the general sectors in the Central Ministries/Departments for the development of Scheduled Castes and Schedules Tribes at least in proportion to their population, both in physical and financial terms. The broad objectives of the TSP are as follows: i) Substantial reduction in poverty and unemployment; ii) Creation of productive assets in favour of Scheduled Tribes to sustain the growth likely to accrue through development efforts; iii) Human resource development of the Scheduled Tribes by providing adequate educational and health services, and iv) Provision of physical and financial security against all types of exploitation and oppression. The Government issued orders that the TSP outlay shall be 5.33% of the total outlay for the year 2016-17, in proportion to the total population of Scheduled Tribes in the state, as per Population Census, 2011. Priority for fund allocation under TSP 2016 is given below:

Table 5: Fund allocation priority under TSP during 2016-17

	Activity	Fund allocation
1.	Education	40-50 %
2.	Health	10-15 %
3.	Agriculture, Horticulture, Animal Husbandry (AH), Fisheries, Dairy & other primary sector	20-30 %
4.	Other income generating schemes to augment Tribal household economy	10-15 %
5.	Administrative structure (including manpower)/ Institutional framework & Research studies	< 5-10 %
	Total	100 %

Research and Development Supports for Sustainable Livelihood Development

Integrated Farming System (IFS)

The ICAR-RC NEH Region has conducted extensive research on developing economically viable sustainable farming systems models on hillslopes to meet the requirements of small and marginal farmers. Watershed based farming system, appropriate soil conservation measures, mixed land use of agri-horti-silvipastoral system, subsidiary source of income through livestock rearing, creation of water harvesting and silt retention structure at lower reaches are the important distinguishing features of the suggested agricultural strategy on the hill slopes. Some of the successful farming system are given below:

i) Agro pastoral farming system

It is suited to the area having slope upto 50 % with more than 1 m soil depth. Contour bunds were made at higher reaches, bench terrace at lower reaches and grass waterways in drainage channel. Top of hillock is kept under forest (fuel-cum-fodder trees, bamboo and timber). Cultivation of crop in hill slope is made in topo-sequence. Rice is grown at the bottom of the hill slope and *cassava*, buckwheat, etc. are grown at upper terraces. Maize is grown next to rice in upper region. Perennial grasses and legumes, such as setaria sphacilata, thin napier, guinea and stylosanthes were planted on terrace raiser. The heights of the grass were maintained at 50-60 cm to avoid shading to crop. This farming system can support 1.18 livestock units (1 unit equal to 1.0 buffalo, 1.25 cattle, 5.0 pigs and 10.0 goats). It reduces soil erosion. A sediment yield from this system was less than 1t ha^{-1} yr^{-1}. Economics of this farming system was estimated as 1.83:1 without integration of livestock and 2.05:1 with the integration of livestock.

ii) Agri-horti-silvi-pastoral farming

It is suited to the area having slope upto 100 % with soil depth greater than 1m. Agricultural land-use towards foot-hills, horticultural in the mid-portion of the hill and silvi-pastoral crops in the top portion of the hill slopes. Contour bunds, bench terrace, grassed water way and half-moon terrace were used. Fodders were grown on the terrace raiser. The produce from this system in 1 ha area can meet the food, fodder and fuel requirement of a tribal family consisting of 5 members. Fodder from this system can feed 10 goats. The benefit: cost ratio of the system was estimated as 2.14:1 and 1.41:1 with and without integration of livestock. Sediment yield from this system was 1.221t ha^{-1} yr^{-1}.

iii) Livestock based farming system

It is suited to the area having slope upto 100 % with shallow soil depth. Contour trenches and grass waterways are used for conserving soil water. Leguminus and non-leguminous annual and perennials, shrubs and trees are planted on the area. It can support 4-5 livestock units(1 unit equal to 1.0 buffalo, 1.25 cattle, 5.0 pigs and 10.0 goats) with setaria and stylo (1:1) mixture of fodder production. It provides substantial income from the farmyard manure and self-sufficiency in the matter of fuel through biogas plants. The benefit cost ratio of the system was 2.08:1. It is expected to retain 90 % of annual rainfall and restricted to soil loss within 2 1t ha^{-1} yr^{-1}.

iv) Horticultural based land-use system

It is suited to the area having slope not more than 100 % with maximum soil depth of 1 m. Contour bunds, half-moon terrace at fruit plant location, grassed waterways and few bench terraces at lower terrace for growing vegetable crops are used in this system. Hilltop is used for forest species to meet fuel and fodder requirement. Yield of newly planted Assam lemon orchard has been found to be 11300, 12800 and 37200 fruits ha^{-1} during third, fourth, and fifth year after planting. Yield of fodder during the early phase are 70-138 t ha^{-1} for bajara x napier hybrids, 44-82 t ha^{-1} for golden timothy grass, 43-74 t ha^{-1} for guinea grass, 45-81 t ha^{-1} for dallies grass and 50-55 t ha^{-1} for maize.

v) Horti-silviculture land-use system

This type of land-use system can be adopted successfully in the area having less than 50 % slope with moderately fertile and deep soils. Tree species, such as *Salix, Populus SP* and *Alnus nepalensis* are grown in the orchard as wind breaker, shelterbelts or filters in order to protect orchard from high velocity wind storms.

vi) Multipurpose tree species and silvi pastoral system

This system can be maintained to non-arable hilly areas with high slope and shallow soil. Forage plant, fodder tree shrubs, grasses and nitrogen fixing plants are planted on the steep slope in the area for afforestation.

Economics of Different Farming Systems

Economic analysis of different micro-watershed based farming systems, namely Dairy farming, Agro-pastoral and Agri-horti-silvo-pastoral systems (Table 6) have shown the economic viability of these systems as an alternative to shifting cultivation. In case of agrihorti-silvipastoral land-use system, about 77 per cent of the net return came from the livestock unit with pigs as a component. From these observations, it could be said that livestock component, particularly pigs, could be important in any farming system which is socially acceptable in the North-Eastern Hill Region. Here it may be mentioned that NEH region accounts for 60 per cent of the total pig population of India and a majority of the tribal people being non-vegetarian, pork forms an important item of their food.

Table 6: Economics of micro-watershed based land-use systems

Economic parameters (Rs/ha)	Dairy farming	Agro - pastoral system	Agri-horti-silvo-pastoral system
1990-91			
Input	23,184	29,675	10,604
Output	37,384	50,882	20,943
Net return	14,200	21,207	10,339
Input : Output ratio	1 : 1.61	1 : 1.71	1 : 1.97
2014-15			
Input			
With livestock	45,030	38,201	11.406
Without livestock	-	17,366	9,197
Output			
With livestock	93,992	78,692	24,468 + 22,275 + 8,750
Without livestock			
Net return			
With livestock	48,960	40,490	13,061 + 22,275 + 8,750
Without livestock		14,575	3,922 + 22,275
Input-Output ratio			
With livestock	1: 2.08	1: 2.05	1: 2.14
Without livestock		1: 1.83	1: 1.42

Source: Anonymous (2015)

It was observed that labour productivity for agricultural land-use system is Rs. 26.98 per man-day, whereas, the same for agri-horti-silvi-pastoral land-use system is Rs. 71.15 per man-day (Table 7). Comparison of the labour productivity of these farming systems with prevailing daily wage rates of agricultural labourers revealed that agri-horti-silvi-pastoral land-use system provided 374 per cent incremental gain over the prevailing wage rate, whereas agricultural land-use system gave only 80 per cent. This implies that a farmer can generate more income by adopting agri-horti-silvi-pastoral land-use system compared to engaging himself as agricultural labour. It is also observed (Table 7) that rate of return to labour cost is 3.65 per ha for agrihorti-silvipastoral land-use system which is much higher than that for the agricultural land-use system (0.80 per ha). The analysis revealed that monthly labour requirement in case of agri-horti-silvi-pastoral system is comparatively uniform varying from 13.11 man-day to 33 man-day per hectare except in July, August and December for which it was 49.03, 8.74 and 9.71 man-day per ha respectively. This implies that agri-horti-silvi-pastoral farming system can provide nearly uniform employment for a person throughout the year except the months of July, August and December. In the months, when labour requirement is less than 30 man-days the person can generate some income by engaging himself in some other activities like working on the farms of fellow farmers or can engage himself in household activities. Another important point under this system is that a farmer with one hectare land needs not to hire labour in any of the months except July, when family members can support him. In case of agricultural land-use system, variation in monthly labour requirement is comparatively high varying from 4 to 97 man-days per ha. This implies that a person opting for this system has to hire labour in the month, when the labour requirement is very high particularly in July, September and November. Therefore, it could be said that agri-horti-silvi-pastoral land use system is superior to other farming systems from labour productivity point of view also.

Table 7: Comparison of labour productivity agricultural and agri-horti silvi-pastoral land-uses

Item	Agriculture land-use	Agri-horti-silvi-pastoral land-use
Labour productivity (Rs/Man-day)	26.98	71.15
Prevailing wage rates for agricultural labourers (Rs/man-day)	15.00	15.00
Incremental gain	11.98 (80.00%)	(374.00%)
Ratio of net return to labour costs	0.80	3.65

Watershed Nanagement Technologies

Watershed is a hydrological unit of an area draining into a common outlet point with more homogeneity of land than other administrative units to ensure coordinated planning on the basis of total available resources, conjunctive use of surface and groundwater, allocating priority for judicious use of water. It is recognized as an ideal unit for planning and development of land, water and vegetation resources. Watershed concept has been used extensively because of importance of water balance in the study of ecosystem. Watershed also allow accurate measurements and monitoring of components of water balance on hydrologic cycle, sediment, energy, heat, carbon and nutrient balances in the watershed ecosystem. This can provide a network of monitoring stations on sites within a basin in a nested form or otherwise to track down the status of pollutants at different points. The monitoring at the level of watersheds or sub-watersheds in a basin will help in analyzing impacts of current and future activities and accordingly plan area specific management options or alternatives based on the priorities as per the intended project objectives.

Therefore, two watershed projects entitled "Hydrological Behaviour of small watershed and sustainability of production systems" under NATP during 1999 to 2004 and "Development of model watersheds in north-eastern region" under NWDPRA during April, 2003 to June, 2009 were undertaken to develop and evaluate the hilly watersheds in eastern Himalayan region. Under the NATP project, watershed from Umsning development block in the Ri-Bhoi district of Meghalaya at a distance of 10 km from the Institute was selected. Under NWDPRA project, three watersheds, namely Wah Umroi watershed in Meghalaya, Sajung watershed in Sikkim and Jalukie–Peren watershed in Nagaland were developed as model watersheds during 2000-2009. On vigorous evaluation, the following resource conserving technologies were identified and recommended for adoption to address the soil and water conservation problems for sustainable agriculture in the region:

Mechanical Measures

Contour/graded bunds

These are small embankments of bunds constructed across the slope to decrease the slope length, which reduces soil erosion and diverts the excess runoff to the designed outlet. The eroded soil is retained within the bunds inter-spaces, which gets leveled up in the course of 3 to 4 years to form bench terrace. If the bunds are constructed with some slope, they are known as graded bunds.

Bench terraces

Bench terraces are series of flat beds constructed across the hill slope separated at regular intervals in a step like formation; spaces between two contours are levelled by cut and fill method. Such measures can be adopted, where the soil depth is more than 1.0 m. It is recommended to convert a steep slope of 20 to 30% to a series of level, or nearly level benches to suit cultivation.

Contour trenches

These are trenches excavated along the contours to break the slope length for reducing the velocity of surface runoff; the water retained in the trenches help in conserving the moisture. It is excavated along contours having slope above 15% to prevent erosion and absorb rain water for horticulture and forestry land.

Half-moon terraces

The half-moon terraces are constructed by cutting land surface in the shape of half-moon to create circular level bed having basin of 1.0-1.5 m diameter on the hill slopes. These are constructed for planting the saplings of fruit crops as well as trees in horticulture and agro-forestry land-use system.

Grassed waterways

Grassed waterways are natural or man-made constructed channels established for the transport of concentrated flow at safe velocities from the catchment using adequate erosion resistant vegetation, which cover the channels. These channels are used for the safe disposal of excess runoff from the crop lands to some safe outlet, namely rivers, reservoirs, streams, etc. without causing soil erosion.

Water harvesting ponds

Dug out cum embankment type of water harvesting structure can be used for creating seasonal and perennial ponds at the foot of the micro-watershed for irrigation and fish farming purposes. Studies on water harvesting in dugout-cum-embankment type of pond in hilly region of North-East India clearly indicate the feasibility of harvesting runoff from watersheds to an extent of 38% of monsoon rainfall. The soil in the area has extremely low water holding capacity and the seepage losses are very high. Thus, water storage may be seasonal or perennial depending on the site condition.

Drainage line treatments

Drainage line treatment structures like drop spillway and retaining wall, rectangular weir, stream water diversion , diversion wall with chute spillway, etc. are constructed on drainage line for safe disposal of excess runoff and arresting soil erosion, arresting of water, diversion of water for irrigation and water resources development.

Impact of Watershed technology

It was recorded that the soil loss from the treated area of main watershed was substantially reduced over the years from 18.9 t ha^{-1} yr^{-1} in 2001 to 6.8 t ha^{-1} yr^{-1} in 2004. The soil loss from the treated sub-watersheds reduced substantially and the values ranged between 7.5 to 9.22 t ha^{-1} yr^{-1}. Due to the drainage line treatments and digging of water harvesting ponds, a substantial portion of silt was trapped in the ponds and channel. It was clear from the values of soil loss from treated sub-watersheds (Upper Shipra and Umroi) that cultivation of tuber crops on high slope induces moderate soil loss with best land management practices.

It was indicated that mixed land-use systems with appropriate soil conservation measures, namely bench terraces, contour trenches, and so forth, were the most effective in retaining 90–100% annual rainfall and simulated the effects of natural forest. It has been observed that the tree based land-use minimized significantly the surface runoff compared to control plots. Hedge row have been found most suitable to reduce the surface runoff followed by alder (*Alnus nepalensis*) and guava (*Psidium guujava*) based land-use systems. For sustainable farming, agri-horti-silvi-pastoral system is one of the best land-use systems with adequate advantage of soil and water conservation. It has been proved that such a system can reduce soil erosion from 42 t ha^{-1} to 1.5 t ha^{-1}, besides providing for rainwater harvesting. This water can be harnessed both for pisciculture as well as irrigation. Agro-horti system of land-use with subsidiary source of income through livestock rearing provides most favourable indication in favour of adopting mixed land-use system as an alternative to shifting cultivation on steep hill sides. Such a system will certainly be technologically feasible, sociologically acceptable, ecologically sound and economically viable.

Hydrological Evaluation of Land-use System

Studies in water based farming systems have shown the scope of using steep slopes for crop production. Soil and water losses from different farming systems were studied in two different locations, which are summarized in Table 8. Some of the potential farming systems, such as agri-horti-silvipastoral systems, etc., have been evaluated at the experimental watersheds for their long term runoff, soil losses and so on. Land-use practices in micro watersheds with soil and water conservation measures were found very effective in retention of rainfall, rainfall retained in-situ varied between 80-100%. The surface runoff from shifting cultivation watershed remained higher in the range of 19.89 mm 54.99 mm in a year compared to other watersheds. Also, as expected the watershed treated with jhum (shifting) cultivation yield the highest peak runoff (86.1 mm/hr) while the one left undisturbed with natural vegetation gave the minimum peak runoff (4.49 mm/hr). However, provision of trenches in fodder based agriculture most effectively conserved moisture and produced peak runoff 7.81 mm/hr). Maximum peak flow rate did not differ in mixed block forest, silvi-pastoral system and agri-horti-silvi-pastoral system showing the effectiveness of suitable soil conservation and appropriate land-use system on hill slopes. The contributions to stream flow in the watersheds having substantial area under natural forest in primarily by subsurface flow (base flow). The watersheds having continuous stream flow characteristics generated base flow to the extent of 70-90% of its total water yield. The highest bade flow was obtained in the pine afforested undisturbed watershed, which constituted on an average 23.5% of annual rainfall.

Table 8: In-situ retention of rainfall under watershed based land use systems

Landuse	Mean slope of the watershed (%)	Soil and water conservation measures	Percent of annual rainfall retained	Soil loss (%) (t ha^{-1} yr^{-1})
Umiam, 1000 m asl, clay loam, 2554 mm annual rainfall**				
Natural fallow	52		Trace – 98.46	0.00-0.05
Agriculture (F)*	32	Contour trench	0.00-98.68	0.00-0.16
Agriculture (C)	32	Contour bunch + Bench terrace	98.53-99.77	Trace -0.33
Agri (C)- horti-silvi-pastoral	32	Contour bunds	98.27-99.99	Trace – 1.22
Horticulture	42	Do	96.18-99.75	Trace- 4.37
Agro-forestry	33	Do	96.83-99.52	Trace-0.38
Forestry	38	-	92.79-98.27	Trace-7.80

F= Fodder based, C= Food crop

Area Treated and Monetary Losses Due to Water Erosion

A total area of 4.06 mha has been treated by spending an amount of Rs.1470.11 crores in the region which accounts for about 7.18 % of the total treated area and 7.55 % of the total expenditure of the country under various resource conservation programmes. About 15.5 % of the TGA of the region has been treated under various programmes against treated area of 17.2% of TGA of the country. Among the states in the development programmes in the country during 1990's, all the previous programmes were converged to develop different areas by adopting participatory watershed management approach. The maximum area has been treated under IWDP, followed by DPAP, NWDPRA and other programmes. In the region, maximum TGA has been treated in the state of Nagaland (35.2%), followed by Sikkim (32.4%), Mizoram (31.4%) and Manipur (19.7%). In the entire region, loss in production of cereals, total pulses, total oil seeds and total of all annual crops accounts for 11.36 %, 1.7 %, 2.5 % and 8.03 % of the production loss in the country, respectively. Monetary loss also follows a similar trend to that of production loss in the entire region. The monetary losses of 10.49 % in cereals, 1.9 % in pulses, 1.69 % in oilseeds and 4.90 % in total of all monetary losses are accounted in the entire region. About 9.9 lakh tons of total production amounting to Rs.6742 million is being lost annually in the region, thus threatening the self-sufficiency of food grains. Therefore, to prevent further land degradation and to increase the agricultural production and income of the community, resource conservation-cum-livelihood programmes need to be implemented on top priority following integrated watershed approach.

Plastic (Agri film) Lining of water Harvesting Pond on Hill slopes (*Jalkund*)

North-eastern region is characterized as highest rainfall zone of the country receiving 2000 to 2400 mm average annual rainfall but suffers from plenty to scarce syndrome. So, mono cropping system is practiced over major area covering hill tops and slopes as moisture is the major limiting factor for sustained crop production. The region is not having sufficient water for irrigation, animal and household purposes. The main reason is that almost all the rainwater is drained out as surface run off (upto 98%) due to terrain and topographical deformation. Most of the time women and children are engaged for collecting and carrying water. Water harvesting by dugout pond is also not safe and economical due to heavy seepage water pressure in hilly slopes. So, micro water harvesting structure like *Jalkund* can increase agricultural and animal production in the region. It is constructed at high ridges of the area. Hence, water harvesting technology in upper reaches of a hill slope using LDPE/HDPE film lined tanks with safety mechanism has been standardized for the Himalayan hills.

Roof top water harvesting

Though North-eastern states are having highest rainfall in the country, the region is still facing water scarcity problem. In hilly areas, most of the rain water is wasted as runoff. During the lean season, agricultural production suffers due to unavailability of irrigation water. Women and children are engaged in carrying water from low land area to house for household consumption. Rainwater which falls on the rooftop is good for use in domestic purpose viz. household consumption, irrigation, etc. By collecting rainwater from rooftop and storing in *Jalkund* or cemented tank or PVC tank, water scarcity problem can be reduced which lead to increase in agricultural and animal production in the villages.

Water harvesting pond

North-eastern region occupies about 80% area on hill slopes & almost all the rainwater is wasted as runoff. During the lean season (Nov.-March), agricultural productions are affected due to unavailability of irrigation water. Runoff water coming from upper hilly areas can be stored by constructing pond at the valley area. Stored water can be retained in lined (cemented, polyethylene) or unlined pond and can effectively be used for supplemental irrigation, animal & fish rearing, duckery & poultry purposes.

Cost effective terrace formation through bio-terracing

Bench terracing although universal acknowledged as an effective soil conservation measure for settled cultivation on the hill slopes is yet to be popularized in the eastern Himalayan region. High cost of construction, decline in soil fertility due to loss of top soil fertility leading to low crop yields in the initial years and often improper shoulder bunds causing devastating soil erosion are the major constrains in creating bench terraces by cut and fill method. But the application of contour hedgerow technology was able to eliminate many of these constraints. A typical design of two system of hedgerow entails the construction of hillside ditches and vegetative barrier by planting of leguminous hedge (*Erythrina indica*) in one system (Bio-terracing-I), and in another system a strong, tugged broom grass (*Thysanolaena maxima*) (Bio-terracing-II), on the bund (rise) of the ditches, which followed contour line of the hillside to create natural terraces through soil reproduction process. The contour hedge rows of both the plant species can function effectively, as soil barriers from the second year of planting and their sufficiency in controlling soil loss beyond the contour bunded area increases as the hedge plant grew with time. The hedge row provides mechanical and biological barrier that help to minimize soil erosion by reducing runoff. A case study on the performance of bio-terracing vis-a-vis cut & fill method has been presented in the table below. During the five years of study, the original slope reduces annually at the rate of 4 % under the bio-terracing I, II and contour bunds without hedge. Thus, a hill slope of 28 % to 38 % can be terraced within 5-7 years through the application of hedge row technology. The bio-terracing is advantageous over cut & fill method of terracing in term of cost effective. Bio-terracing required 75 to 80 man-days as against 497 man-days under cut and fills method (Table 9). Contour hedge row technology provides an option for farming on hill slopes on sustainable basis.

Table 9: Changes in slope and economics of different system of terracing

System of terracing	Original slope (%)	Slope in fifth year	Man-days required/ha (Cost in Rs. ha^{-1})	Crop productivity (kg ha^{-1})			Total biomass (kg ha^{-1})
				Maize	Ginger	Turmeric	
Cut and fill methods of terracing	38.2	-	497	2655	8117	9832	44870
Bio-terracing-I	27.2	7.3	75 (3750)	2674 (1229-3440)	8911	7629	17650
Bio-terracing-II	28.2	8.03	85 (4250)	2250 (1116-2460)	3869	2695	25850
Contour bund without hedge	28.6	8.41	42 (2100)	1610 (1078-2020)	6058	5448	13780

Low cost bamboo frame polyhouse technology for terraced beds in hills.

Bamboo is used for making frame of this type of polyhouse. The shape of roof is arc type and 200 μ thick UV stabilized transparent LDPE film is used as covering material. Bitumen is used to paint on the portion of bamboo which is inside soil. It is done in order to protect bamboo inside soil from termite. The cost of constructing this type of polyhouse is Rs. 140 per m^2 of floor area (at 2006 cost index). It is suitable for growing high value crops like vegetables and flowers during off season. Most profitable cropping sequence in the polyhouse is found to be Capsicum-tomato-lettuce with an income of Rs 97.89 m^{-2} area of land.

Light weight and low cost Plastic Maize Sheller for Drudgery Reduction

Shelling of maize cob manually without any suitable equipment is a time taking and painful job. To overcome this problem, a plastic body hand held type maize sheller was developed by Dr. R.K. Singh and team. It is of tubular shape having diameter 60 mm, length 70 mm and fitted with 4 numbers of fins for shelling of matured maize cob. Tubular body is made of PVC pipe and tapered fins of MS sheet are riveted inside the tubular body. Shelling is done by holding the sheller in one hand and gradually inserting the cob in the tube by other hand with backward and forward twist.

Impact: The plastic maize sheller is rust proof and safe for hands. Shelling capacity is 25-30 kg/h as compared with traditional shelling of 10-12 kg/h. Saves 66% labour and operating time.

Beehive Briquette Mould for Small scale Entrepreneurship by landless Farmers

This equipment is used for making briquettes from charcoal powder. Overall dimension of the equipment is 250 mm × 100 mm. It consists of three parts: a) cylinder, b) base plate with 21 rods and c) cover plate. Base plate has total 21 rods, each of 12 mm diameter and 95 mm height which are welded on it. Cover plate has same number of holes, each of 13 mm diameter so that it can move through the rods on base plate easily. After putting cover plate and cylinder on base plate, the mixture prepared with charcoal powder, clay and water is put into the cylinder and the whole unit is beaten on a wooden plank to increase the compaction of the mixture. Then, the cylinder and cover plate are pulled out of the base plate along with the briquette. It is placed upside down on ground and pressed to release the briquette. Thus, the dimension of each briquette will be 146 mm in diameter and 85 mm in height which perfectly fits in a charcoal/wood stove available in the market.

Impact: It gives an economical way of using charcoal powder which is having low or no market value. Emission of harmful gases, such as CO, CO_2, CH_4, NO and NO_x are in the range of 0.05 % - 0.1 %, 0.1 % - 0.5%, 100-200 ppm, 0.5-3.0 ppm and 0.5-3.5 ppm, respectively which are well within the permissible limit. By making 6000 Beehive Briquettes, a farmer can earn monthly income of Rs. 24000.

Issues and Thrust Areas of Development

Secondary Agriculture for a Sustainable Livelihood Security

The secondary agriculture creates facilities for primary processing, stress management in agriculture, and adds value to the basic agro commodities, which gives opportunity to the farmers to get better returns from their harvest. It also helps in growing rural economy which is entirely based on agriculture. Secondary agriculture may include processing of value added products from fruits and vegetables, extraction of vitamins and minerals from grains, oil from rice bran, starched sugar from corn, milk and protein from soyabean, industrial chemicals and bio-fuel from sugarcane and ligno-cellulosic biomass, making of fiber board from rice straw and high value animal by-products. In addition to that medicinal plants and herbal products which are not yet fully capitalized in NEH region, can add value many-fold to primary agriculture. The region has almost achieved near self-sufficiency in primary agriculture (grains, sugar cane, fruits, vegetables, milk, etc.) and the time has come to pay attention on secondary agriculture.

Government has emphasized the need for value addition in the agricultural sector and has a vision to increase the level of processing from 8 to 20 per cent, value addition from 20 to 35 per cent and share in global food trade from 1.5 per cent to 3 per cent by 2015. To bring this into reality, government has taken initiatives like establishment of mega food parks, creation of integrated cold chains, modernisation of abattoirs, quality control laboratories, human resource development and capacity building through the establishment of National Institute of Food Technology, Entrepreneurship and Management (NIFTEM) and Indian Institute of Crop Processing Technology (IICPT). Government also enacted an integrated food law through Food Safety and Standards Act and set-up Food Safety and Standards Authority of India (FSSAI) to implement the provisions of the Act.

Rural Entrepreneurship – An effective linkage between primary and secondary agriculture

Rural industrialisation can reduce the imbalance between farm and off-farm sector by making use of primary agricultural commodities as the raw material. Promotion of agro-based industries through rural entrepreneurship is one of the effective ways to tackle the problem. It will improve the accessibility to cheaper inputs and farm produce can be utilized for secondary agriculture, which will add value to the primary produce. The inputs necessary for agricultural production can be met by the rural industrialisation which may include of nutrients (micro nutrients and bio-fertilisers), quality seeds and farm mechanisation. Seed production and processing can be taken at a decentralised level, making use of technical backup of Agricultural Universities and private companies. The seed clusters that have emerged in Hyderabad, Bangalore, Nasik and Terai regions can be replicated in north-eastern regions also. Mechanisation is an essential component and can help in moving the high human resource dependent on primary agriculture to other profitable sectors. Some studies have indicated that in the recent years, the farm labour availability has come down due to its absorption in public sector employment generation programme (MGNREGS) and migration to other sectors. There is scope of shifting the excess labour in the 'primary agriculture' to 'secondary agriculture' profitably with proper capacity building. The falling labour productivity can also be tackled through selective mechanisation. The location of food processing industries in rural areas can contribute to the utilization of local production and reduce the lengthy supply chain by establishing direct linkage with producers. This can result in greater price realisation for producers on the one hand and cheaper raw material to the processors on the other. The locally available cheap resources add to the cost effectiveness of processors by improving their supply chain by ensuring stable and remunerative price to the farmers. It is also required to emphasise more on integrated system of food storage with irradiation and cold chain facilities to minimize post-harvest losses and facilitate value addition and create employment opportunities in rural region. The Government needs to provide connectivity through network of roads and uninterrupted power supply to support such industries in rural areas.

The north-eastern hill region of India receives high rainfall of 3000 mm to 5000 mm a year due to its geographical position. The weather of this region is cold and humid, which is highly susceptible for development of micro-organisms to spoil the produces. The region also experiences hot and humid weather, which is highly hazardous from post-harvest management point of view. Due to high rainfall, the sunshine hours are not effective to dry farm produce. The region has still retained conventional ways of production, preservation and processing of food. Due to improper post-harvest management and lack of adequate transportation and storage facilities, a huge quantity of horticultural crops is lost every year. The remoteness and lack of proper communication creates marketing problem and the market prices are usually low due to high arrivals at one place. The post-harvest losses of food grains in this region are also higher than other parts of the country. There are good prospects of processing agricultural produce and by-products in the production catchments of this region using improved post-harvest techniques. This will ensure loss reduction; add value to farm produce, maintain better quality and employment generation, which in turn will contribute to enhancement of skill of rural entrepreneurs and workers.

Due to perishable nature of high value fruits and vegetables, and lack of proper infrastructure and management practices in post-production period, the quantum of loss has been estimated to be very high as per a study conducted by Assam Agricultural University, Jorhat (Table 10 & 11).

Table 10: Post-harvest losses of some horticultural produce

Crop	Post-harvest loss (%)	Approximate monetary loss/ year (Rs. in crores)
Banana	22.00	128.34
Orange	13.95	4.33
Pineapple	9.25	19.33
Cauliflower	15.75	11.03
Tomato	25.25	41.99
Ginger	10.50	11.66

A large quantity of fruits and vegetables (10-25 per cent annually) gets damaged during the process of handling, transportation and marketing (Deka *et al.*, 2004). The highest monetary loss is recorded in banana and lowest in oranges (Table 10). This has reflected the drainage of hard earn revenues due to non-adoption of appropriate post-harvest management and lack of storage facilities in the region.

Table 11: Post-Harvest losses (%) of Horticultural Crops at various stages of handling

Crop	Harvesting	Grading	Transportation	Marketing/Storage		Total
				Wholesaler	Retailer	Loss
Orange	3.25	0.75	1.25	1.20	7.50	13.95
Pineapple	1.73	0.54	1.95	2.66	2.37	9.25
Banana	Nil	Nil	10.00	5.00	7.00	22.00
Ginger	1.5	2.25	1.50	2.75	2.50	10.50
Tomato	0.75	3.75	11.00	2.50	7.25	25.25
Cauliflower	Nil	2.75	7.50	1.75	3.75	15.75
Spine Gourd	Nil	8.30	4.30	1.80	2.70	17.10
Pointed Gourd	Nil	5.40	7.50	1.90	2.10	16.90

Source: Deka *et al.*, 2004

The post-harvest losses vary from crop to crop and stages of post-harvest handling (Table 11). The maximum post-harvest losses were recorded during transportation of tomato and banana, whereas the maximum loss during marketing was recorded in banana (12.00 %) followed by tomato (9.75 %). In order to minimize the post production losses, appropriate secondary agriculture strategies need to be adopted. To raise the farm income, a plan may be chalked out to market some selected high value items like banana, pineapple, citrus, spine gourd, ginger, etc. in different terminal markets of the country. Such a venture may be further linked to enter the export market. This will also help in generating employment avenues for the educated unemployed youths of the region.

The shelf life of Khasi mandarin fruits can be increased by harvesting at optimum maturity and treating with stay fresh wax. Pineapple harvested at appropriate stage and wrapped with shrinkable polyethylene remains fresh for longer duration. Most of the Banana ripens in three to five days after harvesting. To delay the ripening of fruits vacuum packaging or cold room (12°C) storage may be adopted. A number of value added products from jackfruits can be prepared, such as powder, pickle, nectar, jelly, canned bulbs, leather, etc. The shelf life of passion fruits can be extended by treating with liquid paraffin wax and packing in CFB boxes using high-density polyethylene pad. The passion fruit is preserved either by heat processing or freezing. The fruit can be processed to prepare several products like juice, squash, cordial, nectar, juice concentrate, carbonated beverages, etc. The weight loss of ginger rhizomes can be minimized by waxing. The ginger slices of 3-4 mm thickness can be preserved for 8-9 months in acidified brine solution without any deterioration. Several value added products like ginger candy, powder, paste, oil, oleoresin, etc. can be prepared from ginger.

Fruit bruising during harvesting and transport is a major problem. Therefore, the appropriate packaging of fruits and vegetables is very much essential to protect them from injury and water loss. The ultimate goal of packaging must be the easier handling of the produce, better quality and cost effective marketable product. The corrugated fiber board (CFB) boxes can be used for appropriate packaging for most of the fruits and vegetables, especially for transportation. Pineapple fruits in north-east are generally not packed for marketing. The fruits are simply carried in open trucks and sold in the market. Due to this, the loss during transportation is much higher. The suitable storage structures can be used to prolong the shelf life of horticultural and other farm produce. Wherever applicable the low cost evaporative cool chambers at home scale level may be used for short term storage of fruits and vegetables.

Soyabean is a legume crop of short duration and miracle food cultivated extensively in all the uplands in India including the hilly areas. It provides energy, protein, oil and milk products to human beings and fodder for animals in addition to enriching the soil. Techniques have been developed for processing raw soyabean into various value added food products, such as soya milk, soya paneer (tofu), soya fortified biscuits, roasted soya nut, etc. All these value added products can be prepared at home scale level.

Food Processing Industry in North-Eastern Region

Agricultural and horticultural produce based Food Processing Industries can help in reducing post-harvest losses and generating employment, especially in rural areas. At present, the industry accounts for nearly 19% of total industrial outputs, generates 18% of industrial GDP and employs 19% of the industrial labour force. Due to cultivation of wide variety of fruits and vegetables, there has been more emphasis on food processing industries in the north-east region. A number of food processing units have come up in recent past and likely to grow more in near future. There is scope for production and processing of organic food (dried fruits and nuts, processed fruits and vegetables, spices, herbs, oil crops and derived products, etc.) and non-food items, such as cut flowers and pot plants having global market.

Agri-business Entrepreneurship

The scope and opportunities in the agri-business have substantially expended due to globalization of trade and agriculture and the policy reforms at national level, leading to an unprecedented level of corporate interest in this sector. The policies of globalization have taken Indian agriculture into the global village, opening up various opportunities as well as numerous challenges. Agri-business has opened up vast opportunities for value addition, packaging, retailing, and exports of agricultural products with high application of technology and management. Agro-industrial activities benefit a number of groups, including poor and marginalized rural populations and urban agro-industries. Adequate mechanization is the main engine of productivity and competitiveness in agriculture and agro-industries. It allows the transition from subsistence farming to commercial farming. As majority of population in NEH region is dependent on agriculture and this sector also supplies raw material for various industries. Moreover, in places where farming workforce includes children and the elderly, proper tools and machinery can help reduce their workload while maintaining agricultural production. The progressive growth of economy in the region is most likely to be controlled by the agri-business sector.

ICAR Research Complex for NEH Region, Umiam, Meghalaya has established "**Agri-business Incubation Centre** under XIIth Plan Scheme for National Agriculture Innovation Fund (NAIF) with the specific objectives of promoting entrepreneurship in farm tools & implements in the NE region, which is having a low level of mechanization (0.6 kW/ha); converting innovations, and horti & animal based products developed by the Institute into a successful business enterprise; facilitating commercialization of technologies developed by ICAR RC NEH Region; facilitating human resource skill development for agri-business. The specific activities of the ABI includes - Incubation of small and medium enterprise to develop agri-business; acting as Bank-Entrepreneur interface; Commercialization of technologies developed by the Institute; Frontline demonstration of technologies; encouraging and skill development of entrepreneurs, Self Help Groups and Farmers Club to undertake agri-business; Commodity day celebration and organization of Entrepreneurs' meets. The centre also assists in the rehabilitation and restructuring of workshops and factory operations which design and manufacture agro-machinery and tools after incubation. It helps blacksmiths and other providers of rural services to produce and maintain standardized equipment that meets local needs. It facilitates the transfer of innovative technologies to small businesses in the farming and food sectors and consequently greatly improves their competitiveness and market access.

The institute has established and successfully running pilot plants for manufacturing of farm tools & implements, agro-processing centre and meat processing plants. The institute has developed numerous commercially viable technologies and protocol like processing of ginger, turmeric, product development utilizing exotic local fruits like sohiong, chow-chow, pineapple, khasi mandarin, jackfruits and crop varieties, etc. The Institute has state-of-the-art facilities for incubation of technologies. Various technologies were developed and available for commercialization under different categories after providing training and incubation to start up small entreprises. For example, under *small agricultural implements* (4-Row Pre-germinated Paddy Seeder, Plastic Body Maize Sheller, Table Top Rotary Maize Sheller, Manual Zero Till Planter, Adjustable Zero Till Furrow Opener Charcoal Briquette Moulder, etc.) and under *processing technology* (Instant Ginger Candy, Passion fruit and Sohiong fruit Jam and RTS, Minimally Processed Ready-to-Cook Green Jackfruit, Water Soluble powder from Passion fruit and Sohiyong fruit).

General Problems and Constraints Related to Resource Conservation and Livelihood Security.

- Practice of shifting cultivation in hilly slopes of the region causing severe land degradation leading to loss of top fertile soil, forest cover, biodiversity and environmental pollution.
- Lack of appropriate data base related to availability of various natural resources including soil and water in the region which is a major hindrance for the planning.
- Poor socio-economic status leading to low adoption of modern production-cum-conservation practices resulting in high poverty. Rural population of this region accounts for more than 75% and is characterized by fragility, marginality, inaccessibility, cultural heterogeneity and ethnicity.
- Low production and low productivity of crops due to topographical and inaccessibility in the NEH region coupled with use of local varieties, un-scientiûc and faulty methods of cultivation and low fertilizer consumption.
- Soils of this region are acidic to strongly acidic in reaction leading to deficiency and toxicity of essential plant nutrients thereby adversely affecting the crop growth and productivity.
- Poor water resource management and inadequate technological backup for harvesting and utilisation of surface run off for crop production.
- Large scale deforestation, destruction of native germplasm through burning for shifting cultivation leading to loss of biodiversity.

- Deforestation and the resultant loss of soil, especially in the hilly areas, are leading to increased siltation of rivers and streams. The deep pools that are the favoured habitats of many species are rapidly becoming shallow and choked with silt, leading to a decline in habitat. At the same time, swamps, marshes, and other wetlands are increasingly being reclaimed for urban and agricultural expansion.
- Non-availability of adequate and timely agricultural inputs like quality seeds, fertilizers, etc.
- Lack of technical skills and high illiteracy rate and poor management of resources.

  Very limited information on advanced farming and cropping systems suited to different locations.
- Poor infrastructural development like connectivity, education, health, transport, market and other basic facilities in the interior areas of the NEH state leading to poor over all development including agriculture.
- Fragmentation of holdings and right of land ownership issue in the NEH region threatening livelihood security due to poor resource conservation.
- Inadequate agricultural mechanization due to the hilly terrain, lack of investment and mechanized transfer of technology majority of the farmers are below poverty line.
- Loss of interest in agriculture especially among the youth with resultant increase in rural-urban migration as well as preference for non-farm sector.
- The region shall also remain dependent on other states of the country for food which might also record an increase in insurgency related activities.
- Increasing rural poverty due to the non-renumarative agriculture production.
- Lack of policy frame work for channelization of production-processing- market components.
- Large gap between the potential and productivity of major crops. This leads to gap between food production and the requirement at the regional level.

- Sedimentation of hydro-electric/irrigation reservoirs and tanks due to unscientific methods of cultivation.
- The region is much vulnerable to recurrent natural disasters, like drought ûooding, landslides, etc.
- Lack of co-ordination among different developmental/research agencies working in the region leading to poor implementation of resource conservation programmes with low impacts.

References

1. Anonymous, 2015. Annual report, 2014-15. ICAR Research Complex for N.E.H. Region, Umiam, Meghalaya.
2. Anonymous, 2015. Agricultural Statistic at a Glance. Department of Agriculture Co-operation & Farmers' Welfare, Ministry of Agriculture & Farmers' Welfare, Govt. of India.
3. Anonymous, 2016. Ministry of Rural Development. Govt. of India. Available at: http://rural.nic.in/netrural/rural/sites/programmes-schemes.aspx
4. Anonymous, 2016. The financial express: April 27, 2016. Available at: http://www.Financial express.com/economy/mgnrega-scheme-3-n-e-states-on-top-two-at-bottom/ 244193. Accessed on 19/01/2017.
5. Census of India, 2011. Directorate of Census, Ministry of Home Affairs, Government of India, New Delhi.
6. Deka, B.C., Sharma, S., Patgiri, P., Saikia, A., and Hazarika, C. 2004. Post-harvest practices and loss assessment of some commercial horticulture crops of Assam. *Indian Food Packer*. 58 (1): 85-87.
7. Livestock Census, 2012. Department of Animal Husbandry, Dairying and Fisheries, Ministry of Agriculture, Govt. of India.

❏❏❏

Chapter 4

Climate Smart Agriculture in North-Eastern Hill Region: Role of Farmers[1]

Ch. Srinivasa Rao, M. Osman, J.V.N.S. Prasad, K. Ravi Shankar and Prasanna Kumar

ICAR-Central Research Institute for Dryland Agriculture, Santoshnagar, Hyderabad - 500 059, India, Email: director@crida.in

Introduction

The north-eastern hill (NEH) region comprising the states of Arunachal Pradesh, Assam, Manipur, Meghalaya, Mizoram, Nagaland, Tripura and Sikkim has diversified agro-ecosystem and receives good amount of rainfall. Land degradation is a serious problem in this region due to undulating topography while availability of water during off-season is limiting the crop production. The region is highly susceptible to acute soil erosion due to its undulating topography, high intensity rainfall and *Jhum* cultivation.

The extent of soil and nutrient transfer, causing environmental degradation in North-eastern India, has been estimated to be about 601 million tons of soil, and 685.8, 99.8, 511.1, 22.6, 14.0, 57.1, and 43.0 thousand tons of N, P, K, Mn, Zn, Ca, and Mg, respectively. Excessive deforestation coupled with shifting cultivation practices result in significant soil loss (upto 200 t/ha/yr), leading to decline in soil health (Saha *et al.*, 2012). Soil erodibility characteristics under various land-use systems in region indicate that shifting cultivation had the highest erosion ratio (12.46) and soil loss (30.2–170.2 t/ha/yr) followed by conventional agriculture system (10.42 and 5.10–68.20 t/ha/yr, respectively) (Saha *et al.*, 2012). The sediment yield is very high in river basins originating from this area compared to all India average of 15.6 t/ha/year (Sharda *et al.*, 2016).

[1] Paper to be presented in the Conference on Farmers First for Conserving Soil and Water Resources in North-Eastern Region (FFCSWR-2017) to be held at 09-11, 2017, AAU Campus, Guwahati, Assam.

Effect of Climate Change in North-East Region (NER)

The signs of climate change in NER of India are becoming increasingly evident. The annual maximum and mean temperature from 1901 to 2003 has increased by 1.02°C and 0.60°C, respectively. The temperature is projected to rise by another 3–5°C during the latter part of this century (Cline, 2007). The most striking evidence of changes in rainfall comes from significant reduction in amount of rainfall in Cherrapunji, one of the wettest places in the world where annual rainfall has now fallen sharply to less than a third of the quantity that used to be in 1860s. The deficit in rainfall in North-eastern region and the recurrent droughts in the region are adversely impacting agriculture production in the region. The projected decline of rainfall is about +10% during *rabi* season and increase (+20%) during rest of the year in the latter half of this century (Cline, 2007). This indicates a possibility of increasing events of high intensity rainfall in future. Rainfall being intricately linked to crop productivity, these climatic changes will have significant impact on food production in NER, which already has gross deficit in terms of demand and supply.

Despite sharing ~8% of geographical area and ~13% of rainfall, NER contributes only 1.5% to the total food grain production of the country, leading to a perpetual food deficit in the region (Kumar, 2011). This is mostly due to soil fertility and water availability-related constraints attributed mainly to poor management techniques. According to a recent report by the Ministry of Environment and Forests, Government of India, the yield of rice, wheat and maize in NER may further decline upto 35%, 20% and 40%, respectively, in the projected climatic scenario of 2030 (Kumar, 2011). Thus, there is an urgent need to prioritize the development and implementation of various soil and water conservation and management practices, such as *in-situ* conservation, soil erosion control, rainwater harvesting, improvement of irrigation potential, balanced and adequate fertilization, soil organic matter improvement, soil acidity amelioration and conservation agriculture to arrest the resource degradation and improve the soil health and enhance agricultural productivity (Srinivasa Rao *et al.*, 2015).

Farmer's perceptions of climate change in north-east hill ecosystems are rise in temperatures and widespread variability in rainfall. Decreasing amount of rainfall, number of rainy days, mid-season droughts are other perceptions of farmers in this region. Late onset and early withdrawal of rains are the common perceptions reported from Himachal Pradesh (Kimani and Bharadwaj, 2015). Studies on climate change show that rise in temperature will be felt more in higher altitudes (IPCC, 2014; Liu and Chen, 2000). Similarly, in a study in 3 states of India, viz., Andhra Pradesh, Karnataka and Maharashtra (Anantpur, Bijapur, Akola and Solapur), major farmers' perceptions were rise in temperatures and decrease in precipitation. Other perceptions were prolonged dry spells, delayed and shorter rains. The major adaptation measures followed were insurance, change in planting dates, cropping pattern, water harvesting, planting drought resistant crops and their varieties.

Development of Climate Resilient Agriculture in NER

To address the issues related to climate variability and to enhance the adaptive capacity of communities, extensive farmer participatory demonstrations of location-specific climate resilient technologies were initiated on farmers' fields in each of the 121 climatically vulnerable districts of the country as part of National Innovations in Climate Resilient Agriculture (NICRA) under Technology Demonstration Component (TDC) during 2011. The climatically vulnerable districts of the country were identified, based on a scientific analysis of climate related problems; exposure, sensitivity and adaptive capacity in relation to climate change (Rama Rao *et al.*, 2013). The adaptation and mitigation strategies were identified following a bottom-up approach in a participatory mode. The objective of the programme is not only to demonstrate the climate resilient agriculture technologies but also to institutionalize the mechanism at the village level for continued adoption of such practices in a sustainable manner. As part of TDC, demonstrations of climate resilient technologies are being taken up in the NER. The districts involved, the climatic vulnerability addressed and significant interventions are furnished in Table 1 and Figure 1.

Not to scale

Fig. 1: Map showing districts in North-Eastern Region along with climate vulnerabilities

Table 1: Districts involved in TDC of NICRA and the climatic vulnerability addressed

State	District	NICRA village	Climatic vulnerability	Major interventions
Arunachal Pradesh	Tirap	Sipini	Erratic rainfall	Rainwater harvesting, Jalkund Advancement of sowing time for *rabi* crops Soil management
Arunachal Pradesh	West Kameng	Sangti	Drought and cold wave	Mulching for soil and moisture conservation Rainwater harvesting
Arunachal Pradesh	West Siang	Dali, Chisi and Padi	Water stress	Soil and moisture conservation through organic mulching Drought tolerant varieties of rice like Bali and Yamuk
Assam	Sonitpur	Punioni-baghchong	Flood	Integrated farming systems (Rice-fish-poultry) High yielding and short duration Boro rice varieties
Assam	Dibrugarh	Phutahola	Floods and terminal drought	High yielding Boro rice variety Fodder bank
Assam	Dhubri	Udmari	Floods	Short duration rice varieties for post flood Fish management during floods Drainage channels
Assam	Cachar	Salchapra	Floods	Flood tolerant varieties Early short duration Ahu rice variety 'Disang' Introduction of poultry breeds
Manipur	Senapati	Hengbung	Drought, hailstorm and cold wave	Rainwater harvesting, Jalkund Mulching with paddy straw and crop residues Cultivation of short duration Rapeseed var. M 27

Manipur	Imphal East	Chingtha	Flood & Drought	Rainwater harvesting, Farm pond and Jalkund Mushroom cultivation *In-situ* conservation through straw mulching Improved shelter for pig
Meghalaya	Umiam	Kyrdem	Erratic rainfall & cold waves during winter	Rainwater harvesting, Jalkund *In-situ* moisture conservation in ginger
Meghalaya	West Garo Hills-Tura	Marapara	Drought	Rainwater harvesting, Jalkund Zero tillage in green gram Vermi-composting
Mizoram	Lunglei	Hnahthial	Water Stress	Rainwater harvesting, Jalkund Nursery of vegetable seedling
Nagaland	Phek	Thipizumi	Heavy rainfall, frost and landslides	Rainwater harvesting, Jalkund Protected cultivation
Nagaland	Dimapur	Dhansiripar	Delay monsoon	Rainwater harvesting, Jalkund Rock fill dam and irrigation channel
Nagaland	Mokokchung	Aliba	Flood & Drought	Rainwater harvesting through farm pond and well Improved short duration varieties
Sikkim	East Sikkim	Nandok	Hailstorms & drought (winter)	Rainwater harvesting, Jalkund Introduction of HYV's and water stress tolerant varieties of crops
Tripura	West Tripura	North Pullnpur ADC	Water scarcity & cyclones	Water harvesting, Jalkund Vermicomposting for on-farm manure generation Promotion of backyard poultry
Assam	Karbi-Anglong	Sikarighat & Belgaon	Flood/ Drought	Rainwater harvesting structure Submergence tolerant Rice varieties Jalashree & Swarna 1 Integrated Fish cum Duck farming
Manipur	Ukhrul	Ramva	Frost	Rainwater water harvesting structure Deep litter System in backyard poultry (Breed Vanaraja)
Meghalaya	Jaintia Hills	Umjalasiaw	Flood	low cost water harvesting structure *(Jalkund)* *In-situ* moisture conservation practices for vegetables Inter-cropping of broccoli with pea
Mizoram	Serchhip	N.Vanlaiphai	Water scarcity	Water harvesting and recycling *In-situ* moisture conservation practice

				Crop diversification Improved shelters for livestock
Nagaland	Mon	Aboi	Water scarcity	Water harvesting and recycling zero tillage in rapseed Short duration varieties of Soybean, Pearl millet, Maize Introduction of improved breeds for poultry
Tripura	Dhalai	Methrina paschim Kuchainala	Flood	Zero tillage lentil Low cost water harvesting structure System of Rice Intensification Flood tolerant variety Paddy Var. Swarna Sub -1

Role of farmers in development of climate change adaptation is much larger in north-east hill regions as landholdings are small and the terrain is highly undulating. Dependence on family labour is high and the role of women farmer is much larger in various operations, household gardening and livestock management (Srinivasarao *et al.*, 2016a).

To address the climate vulnerabilities in the selected villages, various interventions were planned under four modules; however, the specific intervention under each module for a particular village was need-based and also based on climatic vulnerability and resource endowments of the selected village. The four modules being implemented are: (1) Natural resource management viz., *in-situ* moisture conservation, biomass mulching, residue recycling, manure management, soil health card-based nutrient application, water harvesting and recycling for supplementary or life saving irrigation, improved drainage in high rainfall/flood prone villages, conservation tillage, and water saving irrigation methods. (2) Crop production module consisting of introduction of short-duration and drought/heat-tolerant varieties, modifications in planting dates for post rainy season crops to cope with terminal heat stress, water saving paddy systems viz., System of Rice Intensification (SRI), aerobic, direct seeding, staggered community nursery to meet delay in onset of monsoon, energy-efficient farm machinery through village custom hiring centers for timely completion of farm operations in limited sowing window, location specific inter-cropping systems and suitable agroforestry systems. (3) Module III covers livestock and fisheries interventions through augmentation of feed and fodder production, fodder storage methods, and improved shelter for reducing cold/heat stress in livestock, management of fish ponds/tanks during when water is scarce or surplus and promotion of integrated farming systems as adaptation strategy. (4) Module IV consists of village level institutions for developing enabling mechanisms and support system. This involves collective marketing groups, introduction of weather-based insurance and climate literacy though establishment of automated weather stations. Inputs from the relevant development departments of the state, zonal agricultural research stations of agricultural universities, progressive farmers are being involved in finalizing the interventions and action plans in a participatory manner (Srinivasarao *et al.*, 2016b).

Village Climate Risk Management Committee (VCRMC)

An institutional mechanism was developed in the form of a committee named as Village Climate Risk Management Committee (VCRMC) for enabling implementation of various interventions in a participatory mode towards adaptation to climate change. This committee essentially consists of representatives from village *panchayat*, women farmers, KVK and progressive farmers from 12 to 20 in numbers. Custom Hiring Centres (CHC) were established for timely farm operations and maintenance of farm implements. The revenue generated from the rent collected from hiring of farm implements was made use for maintenance of implements and the surplus amount was deposited in a bank account opened in the name of VCRMC. Similarly, farmers were encouraged to produce seeds of improved varieties of crops and their spread through Village Seed Bank (VSB) and cultivate improved varieties of fodder.

Impacts of Climate Resilient Interventions

Rainwater harvesting in *Jalkunds* was demonstrated, as the region receives significant amount of rainfall and as crops are being affected by the prolonged dry spells. Emphasis was laid on water harvesting and its efficient use, which can be utilized to provide protective irrigation to the crops for successful cultivation which otherwise would lost as runoff.

The KVKs of Zone III under TDC of NICRA facilitated construction of number of low-cost rainwater harvesting structures across the region. The constructed structures had enormous impact on the livelihood of farmers all over the North-Eastern Region as the structures to a great extent reduced the irrigation charges during the dry season.

The interventions followed a participatory approach in which critical inputs were arranged by the KVK. The harvested water was utilized for irrigation of vegetables and to meet the drinking water requirements of livestock. Rejuvenation of existing water bodies was another important measure taken up by the North-Eastern KVKs for enhancing irrigation potential of the region.

In West Garo Hills, the Jalkund of dimension 5 x 4 x 1.5 m with silpaulin as lining material was created for storage of 30 m^3 of harvested water during rainy season which was used for providing critical irrigation during dry spells in winter vegetables, like cabbage, cauliflower and tomato and farmer could realize a total net income of over one lakh during 2014 for a small piece of land (0.8 ha). Similarly, in Chingtha village, Imphal East, 13 farm ponds and 10 Jalkunds were constructed during 2015-16 and 32 numbers of farmers were benefitted from farm pond and 10 farmers from jalkund covering 8 ha area. Harvested water was utilized for vegetable cultivation during *rabi* season. An area of 0.25 ha was irrigated by each farmer by giving one protective irrigation resulting in net income of Rs. 77,992/ ha.

At Nandok village, East Sikkim, about 25 Jalkunds of size 5 x 4 x 1.5 m were created, each having a capacity of 30 m^3. Farmers started cultivating winter crops in a larger area through efficient utilization of harvested water.

Mr. Chhetri, a farmer in Nandok village in East Sikkim cultivated off-season vegetable using harvested water and cultivated cabbage, cauliflower, broccoli, tomato and capsicum. He earned more than Rs. 40,000/ha over a period of three months and also through sale of seedlings @Rs. 1/- per seedling. Due to his efforts, a total of 8.2 ha area was brought under cultivation which was otherwise kept fallow in the village.

Rainwater harvesting in 'Jalkund' and utilization for vegetable cultivation in East Sikkim, Sikkim

Similarly, in Dhansiripar village, Dimapur, 15 *Jalkunds* were constructed having capacity of 30 m^3 and the water was used during dry season for kitchen garden and as a source of drinking water for livestock.

Shri Tashi Lepcha, farmer of Nandok village in East Sikkim, constructed a Jalkund of size 5 x 4 x 1.5 m and covered with a plastic lining of 250 GSM with a potential storage of 30 m^3 of rainwater. The harvested water was utilized by four other farmers in 1.8 ha area as life saving irrigation through micro-irrigation for sole crops, vegetable seedlings and potato. Judicious use of Jalkund water to irrigate *rabi* crops during dry spells with sprinklers made vegetable production a profitable venture. The low cost ponds helped in doubling the cropping intensity with net returns of Rs.1,00,320/ ha.

In Ramva, Ukhrul of Manipur, 5 farm ponds (10 × 30 × 2 m) and 6 Jalkund (5 x 4 x 1.5 m) were constructed for life saving irrigation in the NICRA village. Water was applied through gravitational force by constructing Jalkund at ridge point of catchments. Net return from cabbage cultivation from Jalkund and farm pond was Rs. 3600/ha and Rs. 8400/ha, respectively.

In Hnahthial, Lunglei of Mizoram, 2 RCC structures and 1 farm pond were renovated during 2015-16. Farmers could irrigate their farm during post *kharif* and *rabi* season for cabbage and broccoli in an area of 4.5 ha. The increase in availability of water in winter season resulted in higher production of cabbage (125 q/ha) and broccoli (79 q/ha) with net returns of Rs. 30,345/ha.

Table 2: Summary of rainwater harvesting structures taken up in NICRA villages in NER in 2015-16

State	Districts	Crop	No. of jalkunds	No. of farmers	Area covered (ha)
Arunachal Pradesh	Tirap	Potato, Rapeseed, French bean	6	49	7.5
Manipur	Jainita Hills	Vegetables	1	4	2.0
Manipur	Imphal East	Vegetables	12	21	18.0
Manipur	Ukhrul	Groundnut	10	22	4.4
Meghalaya	Ri Bhoi	Vegetables	12	12	2.8
Meghalaya	West garo hills	Vegetables	5	6	3.5
Mizoram	Lunglei	Vegetables	2	2	0.5
Nagaland	Dimapur	Vegetable	23	23	8.3
Nagaland	Mon	Broccoli, Pea	2	2	0.8
Sikkim	East Sikkim	Vegetables	25	25	13.0
Tripura	Dhalai	Cabbage, Cauliflower	2	2	0.5

Soil erosion causes damage to terraces in hilly areas of Nandok village of East Sikkim. Plantation of fodder grass (Hybrid Napier) on farm bunds was taken up across the slope. Fodder grass helped in trapping the soil during heavy rains and conserved the soil. About 10,000 seedlings were planted in 1 acre, each of 33 farmers in the village.

Planting hybrid napier grass on bunds across slope in East Sikkim, Sikkim

In North Pulinpur ADC village of West Tripura, Furrow Irrigated Raised Bed (FIRB) maize cultivation was demonstrated in 17 farmer's fields covering an area of 3.5 ha. In FIRB method, maize hybrid (Disha-3502) was sown adopting a spacing of 37.5 cm wide beds and 30 cm wide furrow compared to the farmer's practice of line sowing in flat land at a row spacing of 67.5 cm. Life saving irrigation was provided from the water harvesting structure constructed under NICRA. FIRB method gave an additional yield advantage of 8 quintal compared to flat bed method of sowing (33 q/ha).

Paddy is the major crop in the Hnahthial village of Lunglei district. Almost 45% area of the village could not be cultivated due to moisture stress or late onset of monsoon. Demonstration of short duration variety of paddy (CAU-R1) through SRI method was taken up in 10 farmer's field in 8.5 ha area. Besides increase in production, farmers were happy with saving of water while weeds were controlled through cono-weeder. The yield obtained was 38.2 q/ha with net returns of Rs.57625/ha.

In Sipni village of Tirap district, demonstration of high yielding drought tolerant summer rice variety dehangi, which is medium in duration (115 days) with 128 cm height, having medium bold grain type with yield potential of 33.0 q/ha was taken up. The variety is resistant to blast and moderately resistant to stem borer, case worm and leaf folder and tolerant to dry spells. The variety was sown in line by dibbling method during mid-April. The performance of the variety was very satisfactory and out of total 35 ha under summer rice cultivation, the variety was cultivated in 21 ha area in the NICRA village.

Sustainable Intensification in Rice Fallows in the Hilly Regions

Prior to intervention of NICRA, farmers leave fields fallow after rice harvest. ICAR Research Complex for NEH region, demonstrated zero till cultivation of pea in rice fallows. Adoption of raised and sunken bed technique permitted cultivation of other vegetable crops after rice. This technique protected the vegetable crops from excess moisture in the rice fields and gave additional returns from double cropping. Potato (Kufri Giriraj and Kufri Mega), Capsicum (Royal Wonder), Pea (Arkel), Tomato (Badshah) and French Bean (S-9) were cultivated in rice fallows during January to May. Net income from vegetable cultivation ranged from Rs.10,000 to Rs.15,000 from field sizes of 0.1 to 1.5 ha. Benefit cost ratio ranged from 1.8 to 2.5. Similarly in Mizoram, zero till short duration pea (Rachana, 110 days duration) along with mulching with crop residue was demonstrated in NICRA village in lowland rice fallows in 40 ha area on residual soil moisture. The legume crop helped in improving soil nitrogen along with additional returns from the increased cropping intensity.

Farmers of N. Vanlaiphai NICRA village, Serchhip of Mizoram, were not accustomed to cultivate *rabi* crops after the harvest of paddy and keep the land fallow due to scarcity of water. The base flow from shallow open well was collected and lifted for irrigation with the help of pumpset benefitting four farmers covering an area of 1.6 ha. An average yield of 25q/ha was obtained from the farmer field with a net return of Rs. 50,000/ha. An opportunity for supplemental irrigation to field pea during *rabi* season led to additional income generation, which attracted other farmers to cultivate *rabi* crops after harvest of paddy.

Green Manuring and Mulching for *in-situ* Soil Moisture Conservation

Growing of green manure crops increases soil fertility and helps in maintenance of soil moisture status. The district KVK demonstrated *in-situ* moisture conservation with green manuring crop of dhaincha, in rice-rice system in the village Salchapra of Cachar district, Assam. The demonstration was undertaken in 2.0 ha of land benefitting nine farmers in the village to show the potential benefits of the technology.

The potential benefits of *in-situ* soil moisture conservation through green manuring using pulse crops was demonstrated to the farmers by KVK, Tirap, Arunachal Pradesh. Green gram (variety: Pratap) and black gram (variety: T-9) were grown for this purpose. Demonstration of green manuring in 5.0 ha of land was taken up by 10 farmers. Green manuring not only boosted soil moisture conservation but also improved soil health (physical, chemical and biological properties of soil). Farmers were able to reap additional income through harvesting of pods of green gram and black gram and selling those in the local markets.

In Ri-bhoi district, demonstration of straw mulching was taken up in the ginger fields of 20 farmers at Kyrdem village covering an area of 0.56 ha/farmer. In order to impart necessary skills among the ginger growers regarding the technology, a number of training programmes were conducted by the KVK experts in the village. The intervention resulted in realization of almost double yield (259q/ha) of ginger by all the farmers. The horizontal spread of the technology in the village was highly encouraging and effective.

Green manuring dhaincha cultivation in West Garo Hills, Meghalaya

Straw mulching in Ginger fields in Tirap, Arunachal Pradesh

Protected Vegetable Cultivation in Heavy Rainfall Areas

In Kyrdem village, Ri-bhoi, high intensity rainfall, hailstorm, frost and cold wave caused damage to crops during summer and winter season, respectively. A demonstration on low cost polyhouse was conducted in the field of Kong Biona Lymphuid, a progressive farmer from Kyrdem village. The polyhouse was constructed using locally available bamboo for the frame. UV stabilized film of 200 μ (800 gauge) was used for covering the roof and 75% shade net on the sidewalls. The estimated cost of construction of a 100 m^2 size polyhouse varies between Rs. 20,000 to Rs. 25,000 i.e., which works out Rs. 200 to Rs. 250 per m^2. In the past, the nursery of different crops was not raised under protected condition. Thus, the nurseries were under risk of damage from heavy rainfall and hailstorm. Cucumber, capsicum, broccoli, lettuce and tomato were remunerative crops for polyhouse cultivation in this area. The nurseries of crops like tomato, cabbage, cauliflower, chilli, brinjal were raised in polyhouse. Average yield of tomato obtained by the farmers during 2015-16 was 118.63 q/ha with a net return of Rs. 2,66,520 and B: C ratio of 2.28. Similarly, in West Garo Hills, off-season vegetable cultivation was taken up. During the year 2015-16, average yield of 2.7 q/ha of tomato and 0.395 q/ha of capsicum was obtained from 0.1 ha area.

Polyhouse farming helped the farmers in reducing dependency on rainfall and efficient utilization of land and scare water resources. In village Thipuzumi of Phek district, 10 numbers of low cost polyhouse, each measuring 30×10 ft was constructed using locally available low cost materials and polythene. Polythene was provided from the project fund; however, other local materials were contributed by the farmers themselves. Tomato (variety: ruby) was taken up in the polyhouse with support of micro-irrigation system from nearby jalkund. Average yield of tomato realized by the farmers was 118.6 q/ha with a net return of Rs. 2,66,520 and B: C ratio of 2.28. A total number of 30 polyhouses have been constructed in the village.

Poly house vegetable cultivation in Ri-Bhoi, Meghalaya

Polyhouse tomato cultivation in Phek, Nagaland

Summary

Enhancing re-silience of agricultural production to climate change is critical for ensuring food and nutritional security of resource poor small and marginal farmers in North-eastern region of the country. Practices, such as efficient rainwater conservation and its utilization through protective irrigation not only enhanced the cropping intensity but also improved the food and nutritional security. Further, such as *in-situ* conservation measures, soil health management practices, poly-house cultivation, etc., were found to be effective and enhanced the adaptive capacity of farmers to climatic variability. Capacity building of farmers on various aspects related to climate change and demonstrating the climate resilient practices enabled them to effectively face the vagaries of climate and stabilize the productivity.

References

Cline, W. R. 2007. Global Warming and Agriculture: Impact Estimates by Country, Peterson Institute for International Economics, Washington DC, USA.

Inter-governmental Panel on Climate Change, IPCC. 2014. Climate change: impacts, adaptation and vulnerability. Working group II contribution to the IPCC 5th assessment report.

Kimani, N.C. and Bharadwaj, S.K. 2015. Assessment of people's perceptions and adaptations to climate change and variability in mid-hills of Himachal Pradesh, India. International Journal of current microbiology and applied sciences. Vol 4 (8): 47-60 p.

Liu, x. and Chen, B. 2000. Climate warning in the Tibetan Plateau during recent decades. Int. J. Climatology. 20: 1729-1742 p.

Kumar Manoj. 2011. North-East India: Soil and Water Management Imperatives for Food Security in a Changing Climate: Correspondence. Current Science. 101 (9 & 10).

Rama Rao, C.A., Raju, B.M.K., Subba Rao, A.V.M., Rao, K.V., Rao, V.U. M., Ramachandran, K., Venkateswarlu, B. and Sikka, A.K. 2013. Atlas on vulnerability of Indian agriculture to climate change, Central Research Institute for Dryland Agriculture, Hyderabad. 116 p.

Saha, R., Chaudhary, R.S. and Somasundaram, J. 2012. Soil Health Management under Hill Agro-ecosystem of North-East India. Applied and Environmental Soil Science. 9 p.

Sharda, V.N. and Ojasvi, P.R. 2016. A revised soil erosion budget for India: Role of reservoir sedimentation and land-use protection measures. Earth Surf. Process. Landforms. 41: 2007-2023.

Srinivasa Rao, Ch., Lal Rattan, Prasad, J.V.N.S., Gopinath, K.A., Rajbir Singh, Vijay S. Jakkula, Sahrawat, K.L., Venkateswarlu, B., Sikka, A.K. and Virmani, S.M. 2015. Potential and Challenges of Rainfed Farming in India, Advances in Agronomy. 133: 113-181.

Srinivasa Rao, Ch., Gopinath, K. A., Prasad, J.N.V.S., Prasanna Kumar and Singh, A.K. 2016a. Climate Resilient Villages for Sustainable Food Security in Tropical India: Concept, Process, Technologies, Institutions and Impacts. Advances in Agronomy. 140: 101-214.

Srinivasa Rao, Ch., Prasad, J.V.N.S., Osman, M., Prasad, Y.G., Ramana, D.B.V., Srinivas, I., Nagasree, K., Ramarao, CA., Prabhakar, M., Bhaskar, S., Singh, AK., Sikka, AK. And Alagusundaram, K. 2016b. Technology Demonstrations for Enhancing re-silience and adaptive capacity of farmers to climate variability (NICRA-TDC Highlights 2015-16), Central Research Institute for Dryland Agriculture, Hyderabad 154 p.

❑❑❑

Chapter 5

Role of Agroforestry in North-Eastern Hill Region

O.P. Chaturvedi, A.K. Handa and A.R. Uthappa

ICAR-Central Agro-forestry Research Institute, Jhansi-284003 (UP) E-mail : director.cafri@gmail.com

Introduction

Agroforestry is the most natural way of life in the Himalayan region considering the general physiography, vegetation, existing agricultural practices and the attitude of the people. Recognizing the fact the first seminar on Agroforestry was organized at Imphal (Manipur) which paved way for organized agroforestry research in the country. The major benefit that agroforestry can bring to the sloping areas lies in its capacity to combine soil conservation with production function (Dhyani and Chauhan, 1995). The traditional knowledge/ practices and the current status of agroforestry in the region have been well-documented (Dadhwal *et al.*, 1989; Chauhan and Dhyani, 1989; Chauhan *et al.*, 1993; Dhyani and Chauhan, 1995; Singh and Dhyani, 1996). The highlights of these documentations speak of numerous traditional and innovative agroforestry practices adopted by the farmers. Potential agroforestry systems (AFS) identified for different altitudinal ranges, slopes and soil depths involving various trees, shrubs, grasses and crop plants for increased biomass production and environmental protection have been described by Sharda *et al.*, 2001.

The eastern Himalayan region comprising seven sister states and Sikkim, lies between 21.5° and 29.5°N latitude and 88.0° and 97.5° E longitude. In these hills, distinct Agroclimatic conditions exist due to biophysical features, such as high rainfall, acidic soil conditions, preponderance of steep terrain and internal resource heterogeneity leading to fragility of the ecosystems. There is limited opportunity to expand arable land and mechanise them on the hill slopes. Shifting cultivation (jhuming), being a dominant land use practices, provides sustenance to the tribal which has been replaced by an unbalanced and unsustainable form of short cycle jhuming, owing to population pressure and other competing land-use practices. Indicators of sustainability, such as severe soil erosion, depletion of biodiversity, crop productivity and water resources have become prominent (Prasad and Singh, 1994).

Agroforestry Interventions for the Region

The economic output under a production system can be increased as long as, we fully understood current productivity in terms of sustainability and ecological viability. Synthesis of balanced tree/grass/arable crops production systems control runoff, water supply and support soil fertility to maintain the system's sustainability.

Indigenous farming systems particularly on the hill slopes suggest that a diverse product based complex agroforestry have been quickly occurring in jhuming areas of different states.

Until recently, potential of these agroforestry practices to solve land-use problems and enhance productivity was not recognized by research and extension agencies. With the passage of time widespread interest has mounted in implementing agroforestry development programmes. Agroforestry has been popularized among decision-makers as a conservation farming solution to sustain productivity of the fragile hill agro-ecosystems (Garrity, 1995; Singh, 2000a; Singh, 2000b & Singh, 2002). Consequently, there is enormous demand for sound agroforestry technology.

Over the years, land-users in this region have accumulated knowledge of their environment comprising land, water, trees, livestock, etc. This knowledge helped them to evolve many appropriate agroforestry practices tested over long periods and proved beneficial in the management of resources to meet a range of needs from food to shelter (Singh, 1999). What urgently needed is a more quantitative understanding of their extrapolability and agronomic improvements to enhance crop productivity.

Large Cardamom Plantation-An Age-Old Agroforestry System

The indigenous tribes i.e. *lepcha* and *limbu* used to collect large cardamom from natural forest. As these forests possessed to village ownerships, a large cardamom growing in the natural forests were domesticated, it became a plantation crop (Singh *et. al.*, 1989). Farmers evolved a classical tree cutting schedule in plantation area to get continuous supply of fuel wood and fodder without affecting the shade requirement of large cardamom plants. When trees attain about 16 cm basal diameter, they are cut to release young trees growing out of natural regeneration and provide shade to the large cardamom plants. Some innovative farmers grow fodder trees at shade trees. The fodder trees are lopped after harvest of cardamom capsule in November, in such a way that the lopped twigs are not allowed to fall directly on the cardamom plants. Large cardamom plantations comprising of trees and perennial herbs resemble to a natural forest ecosystem. It is a economically viable cash crop with export potential.

Alder Based Agroforestry Systems

Growing agricultural crops in association with alder trees is a remunerative agroforestry system in Nagaland and is practiced in about 22000 ha area of the state. Alder trees enrich soil fertility and the tribal farmers of *Angami, Chakhasang, Chang, Yimchaunger and Konyak* tribes take advantage of that (Gokhle *et al.*, 1985). Alder-based production system is an outstanding example of sustainable land-use that stood the test of time in Khonoma village, Nagaland. In this system, alder seedlings are planted on sloping lands and the trees grow rapidly for six to eight years. At that stage, the trees are pollarded, leaves and twigs burned and the resultant ash mixed with soil to prepare it for raising arable crops. Subsequent pollarding is done once in every 4 to 6 years and the coppices, except 5-6 on top of the main trunk, are cut off and cropping is followed by fallow period of two to four years. The bigger branches and stripped off leaves are used as fuel. Besides fixing atmospheric nitrogen through root nodulation, the litter added to the soil provides phosphorus, potassium, calcium and other nutrients (Sharma and Prasad, 1994, Sharma and Singh, 1994).

Agroforestry for tree fodder production

Fodder trees, mostly native to the place; provide green fodder during the dry winter months from October to March. Tree leaves constitute 25 to 60 per cent of total green fodder fed to livestock in the state like Sikkim (Singh, 1996). Many fodder trees are planted on terrace risers. There is further need to generate information on different tree-crop interactions and to develop techniques of lopping / planting of trees to minimize tree effects on the agricultural productivity.

Different tree species provided fodder at different times of the year. The sustained supply of fodder was turned to the advantage of creating multi-species plantation. It was also suggestive to know number of livestock kept and the way feeding systems were formulated. Lopping commences in October or November and completes in March or April. Generally lopping of evergreen trees is extended upto April and the lopping of deciduous or semi-deciduous trees are completed before the leaf shed. Some of the fodder trees are lopped during May to September, a peak period for agricultural operation in the hills. This is done to save time for agricultural activities. Productivity of trees are being governed by month and intensity of lopping (Singh, 1995). The effect of light, moderate and heavy lopping treatments on a three years old alder (*Alnus glutinosa*) showed that within 24 hours of lopping, transpiration rates decreased and water potential increased and the effects remained for 12 days resulting into enhanced water use efficiency during early periods of re-growth (Singh and Thompson, 1995). Therefore, fodder trees may not tolerate more than 30-40 per cent of leaf area removal to produce fodder on sustainable basis. Among the introduced tree species, *Leucaena leucocephala* variety S 11 does well at low mid-hills (Singh, 1996). It is completely free from leaf spot disease caused by *Camptomeris leucacena*, most common under humid climate of eastern India (Singh, 1995). Leaves of bamboo are also important source of green fodder (Singh, 1999).

Mixed homestead garden

Around the house, a number of trees species viz. arecanut, coconut, tree tomato (*Cryptomandra betacea*), citrus, guava, jackfruit, drumstick (*Moringa spp.*) and some fodder trees are invariably grown with under storey crops, such as vegetable beans, cucurbitis, ginger, turmeric, black pepper and betel vine having largest variation in their combinations from house to house depending upon area available, site conditions (Slope and soil depth) and choice of a family (Singh and Pradhan, 1993). The characteristics of these agroforestry systems can be defined by a range of plant species producing diverse products.

Agri-horti Systems

In low to mid-hills, mandarin is a commercial fruit crop of this region. The fruit orchard is intensively inter-cropped with annual food crops grown on uplands (Singh and Pradhan, 1993).

Tea Plantation

Tea plantation also exists in small pockets upto 1500 m from msl. They are grown without shade trees due to low light intensity and duration of bright sunshine available at the site but wild cherry (*Prunus ceresoides)* is planted as live boarder (Singh and Pradhan, 1993). At foot hills and low hill upto 500 m from msl, there is scope to select shade trees with desirable characteristics for tea production. Shade trees like *Albizzia chinensis* add 2500-5000 kg organic matter per hectare to the soil annually, contained 63-126 kg N 18-36 kg P_2O_5, 22-44 kg K_2O, 32-64 kg CaO and 16-32 kg MgO per hectare. Under shade trees, feeder root of tea develop at a higher rate than without shade and they absorb large amount of harmful infra red radiation from solar spectrum and diverts larger assimilates towards plucking shoot ad the leaf.

Farmers' Innovative Approach

With changing circumstances *Jhum* farmers no longer remained solely dependent on jhuming. Spreading education, opportunity for jobs, improved financial position and resultant decreasing hands in the family to work in *Jhum* fields have positively influenced on land development activity for *pani kheti* in the valley and foot hills, having good forests in the surrounding to ensure perennial source of water springs in Arunachal Pradesh and elsewhere. In jhum field also, a sense of permanent cultivation is being built (Singh, 1996). Now banana and tapioca are being planted on the boundary of jhum field and inside many arable crops are grown. Boundary crops check soil erosion and act as fencing from cattle. Farm forestry has also begun in small way on the jhum field to ensure their right on the land in coming years. Planting of *Terminalia myriocarpa* and *Morus laevigata,* highly valued for timber species, are very common in Arunachal Pradesh.

In Meghalaya, thinned pine tree + ginger and planting broom grass at hill top or in termittently planting in pineapple field or boundary line are worth to be noticed. Department of Agriculture in Mizoram has developed in own contour trench farming for *jhum* areas (Anonymous, 1995).

Development and Management of Agroforestry Systems

Hill agro-ecosystems are generally fragile and can not sustain heavy biotic pressure (Thakur, 1988). Therefore, human bearing capacity of a site should be the guiding factor in any planning process for development.

Silvi-pastoral system in existing forest

Natural forest lands around the villages are potential area for growing shade tolerant forage grasses. Studies on shade tolerance and productivity of grasses under the shade of *Alnus nepalensis* and *Schima wallichii* mixed forest at the mid-hills of Sikkim produced 3.0-5.0 t/ha dry mailer followed by napier x bajra hybrid NB 21 (Rai *et al.,* 1988).

Silvi-pasture systems on wastelands

Silvi-pasture systems in the hills are productive ways to utilize wastelands for extending forage availability during dry winter months through tree leaves and achieve conservation of natural resources, such as land and water. In Sikkim at a mid-altitude, nevaro *(Ficus auriculata)* based silvi-pasture systems have shown production potential of 15.6-51.6 t/ha/annum progressively in subsequent years after planting (Singh *et al.,* 1995). Under the comparable agro-climatic condition of Meghalaya, *Bauhinia* Spp. and *Symingtonia* populnea + Congo signal grass and broom grass systems were found most productive. They had produced edible biomass productivity 23.7 t/ha/year with a car-rying capacity of 25-35 goats (Singh and Satapathy, 1998). These systems can achieve efficient utilization and protection of land resource by sustaining agricultural production through organic recycling or livestock wastes and enhance economic performance of the agroecosys-tems.

Agro-horti-pastoral systems

It has been well-established that food crops viz. maize, rice bean, ginger, turmeric, soyabean and forage crops, such as golden timothy grass, dallis grass can be successfully inter-cropped during first three to four years after establishment of mandarin, guava and avocado orchards with minimal adverse effects on fruit tree (Singh and Pradhan, 1993). Combining production of annual field crops (maize, ground nut, mustard, radish & ginger) and fruit crops (mandarin, lemon, guava and pineapple), livestock (pig & goat), and fuel wood (a short rotation silvi-pasture) in a single land management unit, provided a wholesome strategy to develop a self-sufficient; land-use management system on the hill slope. Agri-horti-silvi-pasture was ideally a model and land-use system. In this system, agriculture had 19.4% of the area (Singh *et al.*, 2005). Out of total agricultural lands, 67.3 per cent (0.1341 ha) area was under terraced land with 173 per cent cropping intensity and 32.7% was under terrace riser planted with guinea grass. The mean productivity of maize, groundnut, mustard, ginger and radish were 2.97, 2.68, 0.48, 14.13 and 32.20 t/ ha year, respectively on terraced lands and 62.50 t/ha/year of forage grass on terrace riser. The organic manure obtained from dung, urine and livestock refuse of goats and pigs, weed biomass and non-edible crop by-products, was sufficient for the system. *In situ* augmented organic manure was applied to terraced lands for crop production at the rate of 20-22 t/ha/year and a part was applied to fishpond. It helped to reduce the requirement of nitrogen fertilizer to one third of the recommended dose and completely met the need of phosphorus and potassium for crop production. Under horticulture, *khasi* mandarin, guava and Assam lemon was planted with an inter-crop of pineapple. Total fruit productivity was 15.1 t/ha/year (Table 7). Moreover, 1.96 t/ha/year tree biomass was built up in fruit trees. Guinea grass inter-planted between fruit trees produced 3.45 t/ha/ year green forage. Under silvi-pasture, mean forage yield obtained from broom grass was 9.72 t/ha/year from winter season feeding of goats. Harvest of first cycle of trees for fuel wood gave sufficient fuel wood (2.14 t/ha/ year) for a standard size family in second cycle of fuel wood plantation, the annual standing tree biomass built up was estimated to be 1.73 t/ha/year. Overall, silvi-pasture had estimated fuel wood yield of 2.01 t/ha/year during the study period. Annual total forage grasses obtained from the micro-watershed was 7.52 t/ ha/year. Livestock (10 pigs and 10 goats) was fed on grain of maize, radish, guinea grass and broom grass and other by-products obtained from annual crops, fruits and silvi-pasture within the micro-watershed, some deficit in individual years was met from purchase out of cash income generated from the sale of livestock and fruits. Annual live weight gain under this system was 583 kg/ha/year. The goat multiplied in number from 10 to 15 in a year. Keeping stocking rate of 10 in number, five adult goats were sold. In family, goats might be either sold or used for home consumption. The system could not meet fully the requirement of cereal rice for a family size of two adults and three children having rice based non-vegetarian diet. However, from the sale of balance of various items of farm produce after home consumption i.e. ginger, radish, fruits, the deficit in the food requirement can be met by purchase from market. Farming system performance kite index of Bimbao *et al.* (1995) indicated that agri-horti-silvi-pasture system had 6.56 times higher biomass productivity, 1.40 times more species diversity, five times more number of bio-resource flows and economically more (Rs.2.94 return, on per rupee investment) efficient than shifting cultivation. The agri-horti-silvi-pasture system as compared to agro-pastoral system had less biomass productivity but richer in species diversity, higher in number of bio-resource flows and also higher in return on per rupee investment. Based on experience gained in this micro-watershed. It is suggested that various components of agri-horti-silvi-pasture should be developed on two hectares of land on a moderate hill slope (35 to 45%) in order to create complete livelihood potential to sustain a standard size family.

Bamboo based agroforestry in water springs areas

Bamboo grooves on water springs and around the farm boundary shade in agricultural lands. Moreover, soil under the shade of bamboo remains saturated during most part of the monsoon. Consequently, crop productivity slides down and many crop fails to grow under this situ-ation. Studies to utilize shaded area under *Bambusa nutans* have revealed that the lands near bamboo groves (11-15 meters from the groves) should be utilized for growing ginger turmeric large cardamom, orchard grasses or Dinanath grass depending upon the choice of grower (Singh *et al.,* 1992). The land beyond this zone may be put to suitable crops of farmer's choice.

Utilization of terrace risers

Terrace risers are sites of high proliferation of weeds. Such risers can be put to perennials, such as Congo signal grass, guinea grass, broom grass, fine stylo or fodder trees, such as *Ficus* spp. to strengthen the efficacy of physical soil conservation measures and achieve the non-competitive land-use strategy for fodder production.

Sericulture based agroforestry systems

Research efforts to develop sericulture based agroforestry at a mid-hill of Meghalaya, is highly relevant for the farming situation of whole region. Dhyani *et al.* (1996) found that major benefit from such a system is its capacity to combine soil conservation with multiple output and income as well as employment generation.

Agroforestry linked with watershed development in Indian Himalayas

Spectacular achievements on sustainable agricultural development through watershed approach at Sukhomajri in the Shiwalik hills, Fakot in the Garhwal Himalayas and Barapani in the NEH region paved the way for launching massive watershed-based programs in the country. Consequently, a large number of integrated resource management projects with donor agencies and their programs, such as EEC, World Bank, DID, NORAD, Indo-German Bilateral Project, IDA/USAID, Indo-Norwegian & Indo-Danish Program, Newzealand and German projects came into being for sustained economic development of the area. The government of India also invested a sizeable amount into many integrated watershed management programmes being implemented by the agriculture, rural development and environmental ministries. National Watershed Development Project for Rainfed Area (NWDPRA) (launched in 1990-91), Integrated Wasteland/Watershed Development Project for Hilly Areas and Watershed Development Fund (HAWDF) in NABARD by Ministry of Agriculture, Drought Prone Area Program (DPAP) and Integrated Wasteland Development Program (IWDP) by Ministry of Rural Development, and National Afforestation and Eco-Development Board by Ministry of Environment and Forests are the Watershed Development projects being implemented in these areas. Invariably, all the programs have included agroforestry in one form or the other as the major activity.

Table 1: Some indigenous and potential agroforestry systems/practices (AFS/P) for the Indian Himalayas and their area of application.

Agro-climatic zone and AFS/P	Brief description	Areas
Temperate		
Agri-silvi-culture Pine trees with field vegetable crops	Khasi pine (*Pinus kesiya*) with pea, radish, potato, sweet potato, cabbage, turnip, cauliflower, mustard or maize.	North-Eastern Hills
Horti-agriculture Plums with vegetables	Plums with pea, radish, cabbage or cauliflower. Trees scattered or on field bunds.	Sikkim, Arunachal Pradesh
Horti-agri-pasture Pears with vege-tables/ broom grass	Pears with cabbage, cauliflower, beans or broom grass	Sikkim
Sub-tropical hill zone **Horti-silvi-culture**		
Alder with large cardamom	Alder (*Alnus nepalensis*) and Chilloni (*Schima wallichii*) with large cardamom	Sikkim, Nagaland, Manipur, Arunahcal Pradesh
Chilloni + pineapple	Chilloni (*Schima wallichii*) with pineapple	Meghalaya, Manipur
Agri-silvi-culture Chilloni with crops	Chilloni (*Schima wallichii)* with ginger/ turmeric	Meghalaya, Manipur
Jhum and *Bun* cultivation	In shifting cultivation areas, selected trees are left to grow during the fallow phase. Crops and vegetables are grown in intimate mixture as intercrops and on raised beds (*buns*).	NEH except Sikkim and Arunachal Pradesh
Agriculture with alder	Alder on field margins with maize, job's tears (*Coix lacryma*), millet, potato and wheat and secondary crops e.g. chillies, pumpkin, taro (*Colocasia*), etc.	Nagaland
Multi-purpose trees on crop lands	MPTS, such as *Prunus cerasoides, A. nepalensis, Ficus, Schima wallichii, Artocarpus, Bambusa, Gmelina arborea, Morus alba*, banana, etc. scattered on bunds, terrace risers or field boundaries	Nagaland, Meghalaya, Arunachal Pradesh, Uttarakhand, J&K, H.P.
Contour hedgerow inter-cropping and Sloping Agric. Land Technology –SALT	*Desmodium rensonni, Tephrosia candida, Indigofera, Leucaena leucocephola, Gliricidia, Robinea, Cassia siamea* and *Crotalaria tetragona* with crops and fruit plants.	Sikkim, Arunachal Pradesh, Nagaland and Uttarakhand
Silvi-pasture Chilloni/pine with broom grass		Nagaland, Meghalaya, Arunachal Pradesh Sikkim

Nevaro with amilloso or Alder	Broom grass is cultivated under the trees of Khasi pine and Chilloni Amiliso (Broom grass) is cultivated with nevaro (*Ficus auriculata*) or utis (Alder)	
Sub-tropical hill and plain zone **Agri- silvi-culture** Apiculture with trees **Horti-agriculture** Sikkim mandarin (*Citrus reticulata*) with field/ vegetable crops) Khasi mandarin with pine-apple/ vegetables	Tussor silkworm rearing on the leaves of oak (*Quercus* spp.) and mulberry plants on terrace risers, CH on contour and muga on Som (*Machilus bombycins*), *Litsea* Inter-crops viz. Maize-wheat; maize+ginger/ buckwheat/millet/pulses/ vegetables/beans/radish/hara simbi/ ricebean; maize+soyabean/ millet; ginger+ ricebean; and maize/sweet potato/ millet/ buckwheat /vegetable beans/radish are grown in the orange orchards in Sikkim Mandarin with pineapple/beans/radish/ ginger/turmeric/cole crops in Meghalaya and other states	Manipur, Meghalaya, Nagaland, Arunahcal Pradesh Sikkim Meghalaya
Sub-tropical plain zone **Tree gardens or Homesteads** 26.Fruit trees in kitchen garden or farm boundaries	Tree tomato (*Cyto maddra betacea),* guava, banana and *Moringa* grown in kitchen garden or farm boundaries.	Meghalaya, Nagaland, Manipur
Mild tropical hill and plain zone **Multi-storeyed** Arecanut based systems Tea with shade trees and betel vine/black	Arecanut+pineapple+betelvine/ black pepper arecanut and betelvine +pineapple and black pepper. All these systems are practised on the most degraded bouldery lands with high slopes and shallow soil depth conditions. Bamboo drip irrigation is used for watering the plants.	Meghalaya, Tripura Arunahcal Pradesh, Manipur, Nagaland, Tripura

pepper **Horti-silvi-pasture** Mixed land-use	Tea cultivation on hill slopes with shade trees e.g. *Alnus nepalensis, Derris robusta*, silver oak, *Albizia* and betel vine or black pepper	Meghalaya, Tripura
Horti-silvi-culture Fruit trees on degraded lands **Agri-silviculture** Plantation crop combination	Arecanut+blackpepper+mandarin; arecanut+ jackfruit+mandarin; bamboo+ broom grass+ *Schima wallichii*; broom grass+tapioca+ sweet potato. These practices are based on the production of fruits, nuts, betel leaf integrated with woody perennials.	Meghalaya, Tripura North-East Hill Region
Horti-silvi-culture Coffee and black pepper with hollock and hollong	Jackfruit+pineapple;mandarin+ jackfruit;arecanut+jackfruit+betel vine	Meghalaya, Arunachal Pradesh
Home gardens	Multipurpose trees such as *Prunus cerasoides*, alder, banana, *Ficus, Schima wallichii, Artocarpus, Gmelina arborea*, mulberry, bamboo, etc. scattered on bunds, terrace risers or field boundaries.	Meghalaya, Arunachal Pradesh
Small production systems	This is a recent introduction. Coffee and black pepper cultivated with hollock (*Terminalia myriopcarpa*) and hollong (*Dipterocarpus macrocarpus*) in forest area and with *Erythrina indica* on crop land.	North-East Hill Region
	Intensive integration of MPTS (*Albizia, Gmelma*, Jackfruit, Banana, Papaya, Coconut, Arecanut) with annual crops and livestock. Silkworm rearing, livestock, honeybee and fish production in and around homesteads.	

Conclusions

With increasing human and livestock population pressure and mounting levels of land degradation, the hill region has become extremely vulnerable. The situation is further compounded by the adverse effects of rising temperature due to increasing atmospheric CO_2 levels. Agroforestry is a promising land-use option capable of ameliorating land degradation and mitigating climate change. Sustainable development and management of these lands through agroforestry innovations not only meets the multiple needs or food, fodder, fuelwood, fibre, fertilizer, etc. The role of agroforestry in addressing the challenges of livelihood, economic development and environmental management needs to be highlighted.

Agroforestry technologies are expected to play a major role in the synthesis of sustainable farming module for economic prosperity of tribals living in north-eastern hills. It may take some more time to make it popular through development agencies. It is essential that expectations about the outcome and output level of agroforestry be moderated and its merit of service functions may economically value in time to come. The degree of success will continue to vary due to its location specificity, appropriate choices and a number of local factors. Much of the conventional wisdom has been challenged by recent research and farming experience in context of adoption of a perfected technology. There is a serious revision on many fronts. It is not certain that technique/technology perfected today will stand the test of experience tomorrow. Hence, a continuum of interactions of research efforts and farm experience should be followed to realize the potential of agroforestry.

References

Anonymous. 1995. *New contour farming system*, Directorate of Agriculture, Aizawal, Govt. of Mizoram.

Bimbao, M.A., Teresita Lopez, Clive Lightfoot. 1995. Learning about sustainability. *ILEIA Newsletter* 11(2): 28-30.

Chauhan, D.S. and Dhyani, S.K. 1989. Traditional Agroforestry Practices in N.E. Himalayan Region of India. Indian J. Dryland Agric. Res. & Dev. 4(2): 73-81.

Chauhan, D.S., Dhyani, S.K. and Desai, A.R. 1993. Productivity potential of grasses in association with *Alnus nepalensis* and pine apple under silvi-horti-pastoral system of agroforestry in Meghalaya. Indian J. Dryland Agric. Res. & Dev. 8:60-64.

Dadhwal, K.S., Narain, P. and Dhyani, S.K. 1989. Agroforestry systems in Garhwal Himalayas of India. *Agroforestry Systems* 7(3): 213"225.

Dhyani, S.K and Chauhan, D.S. 1995. Agroforestry interventions for sustained productivity in North-Eastern Region of India. J. Range Mgmt. & Agroforestry, 16(1): 79-85.

Dhyani, S.K., Chauhan, D.S., Kumar, D., Kushwaha, R.V. and Lepcha, S.T. 1996. Sericulture based agroforestry systems for hilly areas of north-east India. *Agroforestry Systems* 34: 247-58.

Garrity, D.P. 1994. Developing conservation strategies for resources poor farmers in South-east Asia. Paper presented in 8th Int. Soil Conserv. Organization Conf. held from December 4-8, 1994 at New Delhi, India.

Gokhale, A.M., Zeliang, D.K., Kevichasen and Angami, R. 1985. *The use of Alder Trees.* State Council of Education Research and Training, Education Department, Govt. of Nagaland.

Gokhle, A.M., Zeliang, D.K., Kevichusa, R., Angami, T. and Bendangnugsang, S. 1985. The use of alder tree. State Council of Education Research & Training, Govt. of Nagaland, Kohima, India, 20p.

Haridasan, K, Shukla, G.P. and Beniwal, B.S. 1995. *Medicinal Plants of Arunachal Pradesh.* State Forest Research Institute, Govt. of Arunachal Pradesh, Itanagar.

IUCN/UNEP/WWF 1991. Caring for earth- A strategy for sustainable living. Gland, Switzerland pp. 198-201

Prasad R.N., and Singh, K.A. 1994. Integrated land-use management for eastern Himalayan agro-ecosystems. *In: Agroforestry Systems for Sustainable Land Uses* (Eds. Panjab Singh, P.S. Pathak, M.M. Roy). Oxford & IBH Co. Pvt. Ltd., New Delhi. Pp. 228-36.

Rai., R.N., Venugopal, K., Singh, K.A., and Singh, R.D. 1988. Studies on compatibility of grasses under natural forest of utis and chillowane. *Indian J. Hill Farming* IA (1): 71-72.

Sarangi, S.K., Singh, K.A. and R. Singh, 2007. Performance of turmeric (*Curcuma longa*) grown under shade of tree species. *Range Mgmt. & Agroforestry* 28(1): 44-46.

Sharda, V.N., Dhyani, S.K. and Sharma, A.R. 2001. Role of Agroforestry in Disaster Manager. Chapter submitted to HPC (DMP), Deptt. of Agric. & Co-operation, Govt. of India, pp. 33.

Shen, S.Y. and Turbollow, A.F. 1983. Regional Impact of herbaceous production on US agriculture. Paper presented in Symp. On Energy from Biomass and Wastes VII". 24th Jan., 1983. Lake Burnarista, Ela Instt. of Gas Technology, Chicago, USA.

Singh K.A. 2004. Bamboo vegetable farming and agroforestry in *"Agroforestry in 21st century"*, Eds. S. K. Chauhan, S.S. Gill, S. C. Sharma & Rajni Chauhan, Agrotch Publishing Academy, Udaipur. pp. 63-70.

Singh K.A. 2001. North-east palms for productive conservation through agroforestry. *Indian Farming* 51(2): 14-18.

Singh K.A. 2002. Prospects of edible bamboo shoot farming in North-East India. *ICAR NEWS* 8(1): 5-8.

Singh K.A. 2005. Effect of clump density/spacing on the productivity and nutrient uptake in *Bambusa pallida* and the changes in soil properties. *Journal of Bamboo & Rattan* 4(4): 323-334.

Singh K.A. 2006. Nutritive value of edible bamboo species and traditional uses in north-eastern India. *Journal of Bamboo & Rattan* 5(1&2): 29-39.

Singh K.A. 2000a. Biomass, litter fall and changes in the soil properties as influenced by the age and tree density in *Alnus nepalensis. Indian J. agroforestry* 2: 37-42.

Singh K.A. 2000b. Bio-terracing in the North-Eastern hills. *ICAR NEWS* 6(2): 5-6.

Singh K.A. 2002. Application of bio-terracing for land development in North-Eastern Hills of India. *Indian J. Soil Conservation* 30(3): 226-231.

Singh K.A. and Sudhir Kochhar, 2007. Long term productivity potential of selected mutagen induced variants of Bambusa pallida Munro. *Range Mgmt. & Agroforestry* (communicated).

Singh K.A., Prasad, R.N., Satapathy, K.K. and Sharma, U.C. 1993. Need for forage resource development in the hills of Eastern Himalayas- *Indian Farming* 43(8): 21-26.

Singh, K.A. 1999. Resource management and productivity enhancement through agroforestry in the eastern hilly agro-ecosystems of India. *Indian J. agroforestry* 1(1): 63-72.

Singh, K.A., A. Singh and K.K. Satapthy, 2009. Tree based land-use systems for optimization of bio-productivity and resource conservation in a humid subtropical climate. *Range Mgmt. & Agroforestry* 30(1): (in press).

Singh, K.A. 1995. Biomass productivity of multi-purpose tree species on the hill slopes of Barapani, Meghalaya. *J. Hill Research* 8(2): 226-228.

Singh, K.A. 1996. Growth, biomass productivity and coppicing behaviour of multi-purpose tree species at a mid-hill of Sikkim. *Range Mgmt. & Agroforestry* 17(1): 93-97

Singh, K.A. 1996. Effect of age, cutting interval, nitrogen and farmyard manure on root growth parameters of forage grasses in an acid *Inceptisol. Indian J. agric. Sci.* 66 (1): 38-44.

Singh, K.A. 1999. Nutrient contents in tree fodder and the leaves of bamboo used during scarcity period in eastern Himalaya. *Indian J. Anim. Nutrition* 16(2): 178-182.

Singh, K.A. 2006. Potential of forage production under non-competitive land-use in north-eastern hills region. *Indian Farming* 56(5): 22-25.

Singh, K.A. and Kochhar, S. 2005. Studies on forage production, soil binding capacity of grasses and changes in soil properties on the terrace risers. *Range Mgmt. & Agroforestry* 26(2): 113-115.

Singh, K.A. and Pradhan, I.P. 1993. An appraisal of some indigenous farming systems with special reference to agroforestry in Sikkim (Eastern Himalaya). *Indian J. Hill Farming* 6(1): 113-122.

Singh, K.A. and Thompsion, F.B. 1995. Effect of lopping on water potential, transpiration, re-growth, ^{14}C-photosynthate distribution and biomass prediction in *Alnus glutinosa. Tree Physiology* 15: 197-202.

Singh, K.A. Patiram, Singh, L.N. and Rai, R.N. 1992. Effect of bamboo (*Bambusa Nutans* Walls Ex. Munro) shade on the yield of some agricultural crops at the mid-hills of Eastern Himalaya. *Indian Journal of Forestry* 15(4): 339-341.

Singh, K.A., and Satapathy, K.K. 1998. Productivity soil fertility and hydrological behaviour of micro-watershed based silvi-pastoral system on a hill slope. Paper presented in First Int. Agron. Congr. On Agronomy, Environment and Food security for 21st Century from Nov. 23-27, 1998 at New Delhi. Indian Society of Agronomy, IARI, New Delhi.

Singh, K.A., and Yadav, B.P.S. 1995. Advances in forage resources development and its potential for sustained hill agro-ecosystems in eastern Himalaya region. *In: New Visit in Forage Production* (Eds. C.R. Hazra and Bimal Misri). Publication & Information Division Govct. of India.pp. 162-83.

Singh, K.A., Balaraman, N., Bhusan, B. and Pradhan, I.P. 1995. Studies on nevaro (*Ficus auriculata*) based silvo-pastoral system in Eastern Himalaya. *Range Mgmt & Agroforestry* 16 (2): 115-121.

Singh, K.A., Parthasarthy; V.A. and Pradhan, I.P. 1994. Lopped Biomass Prediction Models for Fodder Trees of Eastern Himalaya. IN *"Agroforestry Systems for Degraded Lands", "Proc. Int. Conf. on Sustainable Development of Degraded Lands through Agroforestry in Asia and Pacific",* Eds. Panjab Singh, P.S. Pathak and M.M. Roy, 25-30 November, 1994, organised by Range Management Society of India at IIFT, New Delhi, pp. 435-443.

Singh, K.A., Rai, R.N. Balaraman, N. and Singh, L.N. 1995. Evaluation of forage grasses for the mid-hills of Sikkim (Eastern Himalaya). *Range Mgmt. & Agroforestry* 16 (1): 39-45.

Singh, K.A., Rai, R.N., Patiram and Bhutia, D.T. 1989. Large cardamom (*Amomum subulatum Roxb.)* plantation-an age-old agroforestry system in Eastern Himalayas. *Agroforestry Systems* 9: 241-257.

Singh, K.A., Yadav, B.P.S. and Goswami, S.N. 1996. Forage Resource Management for Sustained Livestock Productivity in *Himalayan Agro-Ecosystems* Published by Director, ICAR Research Complex for NEH Region, Barapani, Meghalaya, 93p.

Sonowal, D.K., Tripathi, A.K. and Kirekha, V. 1989. Zabo-an indigenous farming system of Nagaland. Indian J. Hill Farming 2:1-8.

Sundriyal, M. and Sundriyal, R.C. 1998. Wild plants of food in Sikkim Himalaya" Aspects of diversity and utilization. *In: Perspectives for Planning and Development in North-East India* (Ed. R.C. Sundriyal). G.B. Pant Institute of Himalayan Development, HIMVIKAS Occasional Publication No.1.pp. 138-49.

Thakur, M.R. 1988. Degradation of Himalayan environment and its conservation strategies. *In: Proc. Nat. Symp. On Role of Science and Technology in Environmental Preservation with Special Ref. to Hill Development*. Held from February 15-17, 1988 at Sikkim. Sikkim Science Society, Gangtok. Pp. 1-17.

❑❑❑

Chapter 6

Organic Farming for Hill Farmers: Perspective, Technological Options, Prospects and Constraints

R.K. Avasthe*, Ashish Yadav, Subhash Babu[1], Raghavendra Singh and Matber Singh

* ICAR-National Organic Farming Research Institute, Tadong-737102, Gangtok, Sikkim, E-mail: ravisikkim@yahoo.co.in, [1]ICAR Research Complex for NEH Region, Umiam-793 103, Meghalaya

ABSTRACT

Organic farming is holistic food production system, which promotes and enhances agro-ecosystem health including quality and healthy food. There are generally two types of organic farming depending on the purpose. One is aimed at increasing efficiency of the large scale production of crops and focuses on broader based market, which also includes foreign market using organic certification system. This is extremely market and profit oriented, whereas the other focuses on food safety, environment consciousness and local sustainability. There is need to promote both types of organic farming, however, more emphasis should be given to traditional crops of the region. The total area under organic certification is 5.71 million hectare (2015-16) in India. This includes 26% cultivable area with 1.49 million hectare and rest 74% (4.22 million hectare) forest and wild area for collection of minor forest produces. Organic farming in hills offers the most sustainable solution for developing the agricultural sector and provides food security with least negative impacts on the environment and also offers solutions for sound rural development, provides healthy food and creates jobs. Soil and soil health is one of the basic principles of organic cultivation. The major challenge in organic farming is the availability of huge quantities of organic inputs for satisfying the farm demand. Use of animal excreta based manure alone is not sufficient for meeting the nutrient needs of the crops. It is therefore, necessary to utilize all the sources of nutrients available on and off farm effectively.

Introduction

The term 'Organic' means origin from a living thing and farming with the philosophy of organic is to make production system alive with long life. Organic farming is holistic food production system, which promotes and enhances agro-ecosystem health including quality and healthy food. Organic farming aimed at a healthy environment, healthy food and healthy consumers. It has very broad meanings and uses natural and eco-friendly methods and materials in agriculture. No chemical fertilizer and pesticides or weed-killers are used. There are generally two types of organic farming depending on the purpose. One is aimed at increasing efficiency of the large scale production of crops and focuses on broader based market which also includes foreign market using organic certification system. This is extremely market and profit-oriented, whereas the other focuses on food safety, environment consciousness and local sustainability. There is need to promote both types of organic farming, however, more emphasis should be given to traditional crops of the region.

Presently people are more health conscious and demand healthy food products. Organic food helps in improving the immune system because it contains the toxic free necessary nutrients which the body needs to be healthy. Because of increased consumer education and awareness, there is a movement towards organic produce, which are not treated with the detrimental inputs. Moreover, the environmental destruction by the chemical industrial model of agriculture necessitates a shift towards agro-ecological forms of production. Organic producers only use inputs and materials which are of organic origin. A variety of crops or plants are planted together to diversify the production system.

Basic Concept in Organic Farming

- Improvement in the biological properties of soil is the main focus in organic farming.
- Conservation of energy and resources in production system.
- Control of pests, diseases, and weeds is achieved largely by the development of an ecological balance within the system and by

the use of bio-pesticides, bio-control agents and adopting good agricultural practices.

- Recycling of all the wastes and manures within a farm.

Soil and soil health is one of the basic principles of organic cultivation. If the soil is healthy, plants are not stressed and will considerably reduce the amount of diseases, fewer infestations by pests and weeds will also be less problematic. It is important that the organic producer understands the soil and working of plants to obtain the best results. Conservation of bio-diversity and soil health contributes to environmental friendly practices and helps to lessen the negative effect caused by climate change. Traditional agricultural and horticultural production systems are inefficient resulting in low yields, poor product quality and low profits. The crop diversification in agriculture will improve land-use, generate employment and provide nutritional security to people. To sustain the production and productivity of diverse organic crops, eco-friendly organic nutrient management system is required.

The International Federation of Organic Agriculture Movements (IFOAM) has formulated four broad principles of organic farming, which are the basic roots for organic agriculture growth and development in a global context. They are:

1. ***Principle of Health:*** Organic agriculture should sustain and enhance the health of soil, plant, animal, human and planet as one and indivisible. Health is the wholeness and integrity of living systems. It is not simply the absence of illness, but the maintenance of physical, mental, social and ecological well-being.

2. ***Principle of Ecology:*** Organic agriculture should attain ecological balance through the design of farming systems, establishment of habitats and maintenance of genetic and agricultural diversity. Those who produce, process, trade, or consume organic products should protect and benefit the common environment including landscapes, climate, habitats, biodiversity, air and water.

3. ***Principle of Fairness:*** Organic agriculture should build on relationships that ensure fairness with regard to the common environment and life opportunities. Fairness is characterized by equity, respect, justice and stewardship of the shared world, both among people and in their relations to other living beings.

4. ***Principles of Care:*** Organic agriculture should be managed in a precautionary and responsible manner to protect the health and well-being of current and future generations and the environment. It should prevent significant risks by adopting appropriate technologies and rejecting unpredictable ones, such as genetic engineering.

Key Characteristics of Organic Farming

- Protecting the long term fertility of soils by maintaining organic matter levels, encouraging soil biological activity, and careful mechanical intervention.
- Providing crop nutrients indirectly using relatively insoluble nutrient sources which are made available to the plant by the action of soil micro-organisms.
- Nitrogen self-sufficiency through the use of legumes and biological nitrogen fixation, as well as effective recycling of organic materials including crop residues and livestock manures.
- Weed, disease and pest control relying primarily on crop rotations, natural predators, diversity, organic manuring, resistant varieties and limited (preferably minimal) thermal, biological and chemical intervention.
- The extensive management of livestock, paying full regard to their evolutionary adaptations, behavioural needs and animal welfare issues with respect to nutrition, housing, health, breeding and rearing.
- Careful attention to the impact of the farming system on the wider environment and the conservation of wildlife and natural habitats.

How Do I Become an Organic Producer?

Producers wanting to become organic growers or producers must have knowledge of the system and related requirements and regulations. The system refers to the following factors:

1. What is expected from an organic farmer?
2. Which inputs are allowed and which are prohibited in organic agriculture?
3. The production methods used in organic agriculture; why and how it is used?

4. How to register as a certified organic farmer?
5. How to market organic produce to get the utmost advantage for the producer?

The organic producer has to obtain a set of the organic regulations from the certifying body they want to register with. The aim in this regard is to supply the producer with the same information as is given in the regulations. The regulations can also be obtained from the organic certification body. The decision is with the producer to register as organic or just to produce their products organically. If the producer is only an organic producer, without being certified, he or she cannot sell the products as certified organic and get a possible advantage of a higher price. Certification or registration takes place through one of the organic certification bodies. An annual fee is payable to the certification body and an inspector is sent out to the farm to ensure the producer adheres to organic regulations. Inspections take place annually. Certified organic producers receive a membership number and logo of the certification agency, which they have to show on the products to sell it as certified organic. Producers are then legally certified as organic producers and have to adhere to the organic regulations to remain certified organic producers. The entire farm of a producer can be certified, or a portion or any number of crops. There has to be a clear distinction between certified organic crops and conventional crops. This remains one of the areas that small scale farmers and consumers have to challenge, particularly the cost involved in certification. In India, mostly the Participatory Gaurantee System (PGS) is used, where farmers guarantee or certify the agro-ecological or organic status of their produce.

Requirements for Producers to Fulfill Before Certification

- Complete a detailed form about the cultivation methods and inputs used on the production unit or farm.
- Keep complete records of the production history of the farm, with maps included - which crops were produced on which fields and when.
- Complete sets of documents are required to show which crops were produced on which fields and the amount of inputs used for the fields. It is quite easy to do it on a map. Inputs include compost and manure.

- No disallowed chemicals should be used – that is fertilizer, pest control and weed-killers.
- A transition (conversion) period at the beginning of the certification is applicable - two to three years, depending on the crop being cultivated.
- The certification authority will send an organic inspector to the farm to gain information about the cultivation system on the production unit and to check the records.
- A certification committee (from the certification authority) will make a decision whether the farm can be certified organic and make recommendations regarding changes and improvements.
- Audits or inspections take place annually and an annual licence and inspection fee is payable. This will allow the producer to use the organic emblem.

Status of Organic Farming in The World

Currently, 43.7 million hectares area (0.99% of total agricultural land) is under organic agricultural management worldwide (FiBL and IFOAM, 2016). The regions with the largest areas of organic agricultural land is Oceania (17.3 million hectares, 39.7 per cent of the world's organic agricultural land) followed by Europe (11.6 million hectares, 26.6 per cent), Latin America has 6.8 million hectares (15.5 per cent) followed by Asia (3.6 million hectares, 8.2 per cent), North America (3.1 million hectares, 7.1 per cent) and Africa (1.3 million hectares, 2.9 per cent) (FiBL and IFOAM, 2016). Apart from the organic agricultural land, there are further organic areas, such as non-timber forest product collection areas (wild areas). The largest part of these wild collection areas are for bee-keeping. Other areas are used for aquaculture; some are forest or grazing areas on non-agricultural land. These areas totaled 37.55 million hectares, and all the organic areas together summed up to 81.2 million hectares. The countries with the largest market for organic food are the United States (27.1 Billion euros), followed by Germany (7.9 billion euros), France (4.8 billion euros), and China (3.7 billion euros). The highest per capita consumption is in European countries. In 2014, Switzerland had the highest per capita consumption (221 euros) worldwide, followed by Luxembourg (164 euros), and Denmark (162 euros) (FiBL and IFOAM, 2016).

Status of Organic Farming in India

As per the available statistics, India's rank in terms of world's organic agricultural land was 15 as per 2013 data (FIBL & IFOAM, 2016). The total area under organic certification is 5.71 million hectare (2015-16). This includes 26% cultivable area with 1.49 million hectare and rest 74% (4.22 million hectare) forest and wild area for collection of minor forest produces (www.apeda.gov.in). Area under organic management in India is continuously increasing. India produced around 1.35 million tonnes (2015-16) of certified organic products. During 2012-13, India exported 1,65,262 MT of 135 varieties of organic products valued at US$ 534 million (INR 3300 crores). In the same period, the domestic market grew annually at 15-25%. A survey conducted by ICCOA (www.iccoa.org) found the domestic market in 2012-13 to be worth INR 600 crores. Tea is currently the most common product produced by means of organic farming in India. The most common organic products being exported from India are organic tea and spices. Vegetables and fruits are the major organic products desired by the Indian customers.

Sustainable Technological Options for Hill Farmers

Organic farming in hills offers the most sustainable solution for developing the agricultural sector and provides food security with least negative impacts on the environment and also offers solutions for sound rural development, provides healthy food and creates jobs. Furthermore, organic farming reduces the vulnerability of the farmers to climate change and variability by comprising highly diverse farming systems and thus, increases the diversity of income sources and the flexibility to cope with adverse effects of climate change and variability, such as changed rainfall patterns. This leads to higher economic and ecological stability through optimized ecological balance and risk-spreading and by low-risk farming strategy with reduced input costs and, therefore, lower risks with partial or total crop failure due to extreme weather events or changed conditions in the wake of climate change and variability. The demand for organic produce is very high, hence cost of organic products is 2–3 times higher than the conventional ones, depending on the products and area. Organic farming avoids nutrient exploitation and increases soil organic matter content. In consequence, soils under organic farming capture and store more water than soils under conventional cultivation. Crops and crop varieties used in organic farming are usually well-adapted to the local environment. Technologies for sustainable organic farming cover the whole spectrum of farming systems. All farming systems, from natural farming to organic farming, have the potential to be locally sustainable. Whether they are in practice depends on farmers adopting the appropriate technology and management practices in the specific agro-ecological environment within the right policy framework. There is no unique system that can be identified as sustainable and no single path to sustainability. There can be a co-existence of more-intensive farming system with more-extensive systems that overall provide environmental benefits, while meeting demands for food. Some technological options for hill farmers are as follows:

a) Mixed crop-livestock production/integrated farming

It describes farming in which livestock and crops are produced within a co-ordinated framework, according to the principles that farm operations should be linked to create closed loops, including the internal use of crop products for feed, management of farm waste for fertilization, and diversification of farm production. Approaches may include the integration of forage crops into crop rotations, and organic nutrient management combining livestock manure, compost, green manures and bio-fertilizers. Integrated crop-livestock systems are most suited to medium intensive mixed-farming systems. In each case, the principles of these production systems have to be adapted to local environmental and economic conditions. Under present agricultural scenario, reducing landholding and declining factor productivity, integrated farming system appears to be the best option for enhancing the land productivity and ensuring livelihood security of small and marginal farmers.

b) Innovative production systems for hill agriculture

System of Rice Intensification (SRI) provides a new approach to rice cultivation, which is less water intensive, saves paddy seeds and requires lower application of chemicals and fertilizers and performs extremely well under organic management. SRI has major advantages over the conventional paddy cultivation. It requires only half the water per hectare compared to conventional farming practices. This is extremely important in view of the water crisis facing the country. Increased productivity leads to increase in farmers' income.

c) Diversified farming systems

It may define as a system of agricultural production that, through a range of practices, incorporates agro-biodiversity across multiple spatial and/or temporal scales. Diversified farming systems share much in common with organic, multi-functional, sustainable, and agro-ecological management approaches and outcomes. The key indicators of a diversified farming system is that diversification across ecological, spatial and temporal scales serves as the mechanism for maintaining and regenerating the biotic interactions and in turn, the ecosystem services- *e.g.*, soil quality, nitrogen fixation, pollination, and pest control- that provide critical inputs to agriculture.

d) Diversifying crop rotations

It includes crop rotations with or without legumes. The use of rotations with legumes brings particular benefits for nitrogen fixation, and legume crops often host more diverse and beneficial invertebrate populations, such as pollinators. Crop rotation including legumes and other green manuring crops should be practiced. Cover crops, catch crops and mulching should be done for conserving soil fertility and reduce soil loss in hilly area. Sufficient diversification should be obtained to take care of the pest and disease pressure and to improve soil fertility, microbial activity and general soil health. Legumes rotation can fix atmospheric N_2 to an extent of 135-488 kg ha^{-1}.

e) Restricting agricultural activities on slopes/contour farming

It includes a ban on the growing of row crops, such as maize, potatoes, sugar beet, and sunflowers on slopes above a specific gradient. Contour farming is the alignment of soil activities (ploughing, furrowing, planting, *etc.*) with contours in order to slow soil erosion and increase water infiltration.

f) Crop residue management

In-field practices, such as incorporation of straw in soil. When the crop residues are retained on soil surface in combination with no tillage, it initiates processes that lead to improved soil quality and overall resource enhancement. Benefits of crop residues retention and management have been demonstrated through its large-scale adoption in many socio-economic and agro-ecological situations in different countries the world over. Since limited application (due to low crop response) of K leads to an accelerated depletion of other nutrients in addition to K, crop residue incorporation (with high in K) can alleviate these problems here, besides improving long-term nutrient balances of cations and restoring Soil Organic Carbon. Recently, it is shown that 2-4 mg N could be fixed by heterotrophic micro-organism per gram of straw added in flooded soil.

g) Agroforestry

Agroforestry is an integrated land-use system that combines cultivation of trees and shrubs with annual crops and/or livestock on the same land, with the aim of benefitting from the complementarities. It is suitable for extensive farming in less favoured areas as well as semi-subsistence farming, and not for large-scale crop and corporate farming. It has the potential to improve livelihood, as it offers multiple alternatives and opportunities to farmers to improve farm production and incomes and also provide productive and protective forest functions to the ecosystems while protecting the natural environment across the globe. The multi-purpose trees in agroforestry systems play a major role in organic agriculture production by supporting food production through increasing the soil's ability to support organic agriculture. A large number of forest and fruit producing trees are integral parts of traditional homestead and other agroforestry farming systems. In hills for sustainable development of agroforestry systems, the maintenance of traditional agroforestry systems and creation of new systems is required by designing context-specific silvi-cultural and farming systems to optimize food production, carbon sequestration and biodiversity conservation. Agroforestry systems are believed to have a higher potential for carbon sequestration than pastures or field crops. In addition to sequestering C in biomass and soil, these systems can contribute to both carbon conservation and carbon substitution.

h) Extensive pasture management

In situations where this management option is planned for carbon sequestration and adaptation benefits, it should focus on adjusting grazing rates and introducing rotational grazing to avoid soil erosion and optimize vegetation growth; where it is planned for biodiversity benefits, it may require greater decreases in grazing rates and conversion to mixed livestock grazing (*e.g.* sheep and horses), or more intensive grazing rates for certain types of grasslands.

i) Use a wider range of livestock breeds

This option conserves livestock genetic diversity but also brings benefits for grassland management through the use of breeds that are adapted to grazing on rough forage in all seasons, including grazing in particularly wet or dry conditions.

j) Biogas technology

Biogas is the product of anaerobic digestion using organic matters, slurry, food waste and other green residue by methanogenic bacteria. Biogas qualifies on the merits that this technology utilizes organic agricultural waste and converts it to fuel and fertilizer. Direct impacts of biogas are fuel-wood, agriculture residue, livestock manure, and kerosene savings. Biogas slurry is rich in nutrients leading to substantial increases in soil fertility and crop production has also been observed. Biogas also solves the problem of indoor air pollution and improves household or communal sanitation.

k) Farm mechanization

With advances in mechanization, farmers can more efficiently tend to their crops and produce more with less manpower. In modern times, powered machinery has replaced many jobs, formerly carried out by working animals, such as oxen, horses and mules.

l) Use of information and communication technology (ICT)

Organizations are using ICT to reduce carbon emissions and develop and support business models with a green focus. It is believed that the use of ICT improves energy efficiency in the economy, starting with buildings, lighting and the power grid. ICT enables economy a green behaviour.

m) Mobile technology

Mobile technology can enable farmers to increase their yields by connecting them through text messages and helplines to agricultural market information, best practices, and extension services designed to meet their localized needs.

The major challenge in organic farming is the availability of huge quantities of organic inputs for satisfying the farm demand. Use of animal excreta based manure alone is not sufficient for meeting the nutrient needs of the crops. It is, therefore, necessary to utilize all the sources available on and off farm effectively. The resource components available for nutrient management in organic agriculture are: farmyard manure, crop residue, weed biomass, green manures, biofertilizers, composts/phospho-compost, vermicomposting, oilcakes, mulching / cover crop, liquid manures, biodynamic preparation, botanicals, legumes in cropping sequence and certified commercial products. The major nutrient content of different organic product and micro-nutrient content of different organic products are presented in Table 1 and 2.

Table 1: Major nutrient content of different organic products

Sources	Nitrogen (%)	Phosphorus (%)	Potassium (%)
FYM	0.93	0.36	0.92
Vermicompost	2.0	1.0	2.0
Pig manure	1.19	0.38	0.98
Poultry manure	1.82	0.51	2.10
Cattle urine	1.2	-	1.2
Crotolaria juncea	3.5	0.33	2.38
Tephrosia purpurea	3.11	0.23	1.24
Eupatorium odoratum	2.38	0.07	2.84
Ambrossia artimisifolia	3.19	0.22	4.38
Rape seed cake	4.8	2.0	1.3
Neem cake	5.2	1.1	1.5

Table 2: Micro-nutrient content of different organic product (mg kg^{-1})

Sources	Iron (Fe)	Copper (Cu)	Zinc (Zn)	Manganese (Mn)	Boron (B)	Molybdenum (Mo)
FYM	2600	2.5	57	250	2.1	0.13
Compost	-	450	9.4	12.4	5.8	0.10
Pig manure	1200	8.9	50	70	-	-
Poultry manure	1400	7.1	90	210	5.0	-
Goat manure	-	61	2570	150	4600	-

Eco-Friendly Nutrient Management Practices

Maintenance of soil fertility may be achieved through organic matter recycling, enrichment of compost, vermicomposting, animal manures, urine, farmyard manure, litter composting, use of botanicals, green manuring, *etc.* Use of bio-fertilizers like *Azolla, Azospirillum, Azotobacter, Rhizobium* culture, PSB, *etc.* can be used. Saw dust from untreated wood, calcified seaweed, limestone, gypsum, chalk, magnesium rock and rock phosphate can be used. Various sprays like vermiwash and liquid manure, *etc.* can be used in crops for nourishing the soil and plant.

Manures

Manures are plant and animal wastes that are used as sources of plant nutrients. They release nutrients after their decomposition. Manures can be grouped into bulky organic manures and concentrated organic manures based on concentration of the nutrients.

Bulky Organic Manures: Bulky organic manures contain small percentage of nutrients and they are applied in large quantities. Farmyard manure (FYM), compost and green manure are the most important and widely used bulky organic manures.

Farmyard Manure: Farmyard manure refers to the decomposed mixture of dung and urine of farm animals along with litter and left over material from roughages or fodder fed to the cattle. On an average well decomposed farmyard manure contains 0.93 per cent N, 0.36 per cent P_2O_5 and 0.92 per cent K_2O. Urine, which is wasted, contains one per cent nitrogen and 1.35 per cent potassium. If urine is not collected in the bedding, it can be collected along with washing of the cattleshed in a cemented pit, from which it is later added to the farmyard manure pit. Chemical preservatives gypsum is spread in the cattleshed which absorbs urine and prevents volatilization loss of urea present in the urine and also adds calcium and sulphur. Superphosphate also acts similarly in reducing losses and also increases phosphorus content.

Partially Rotten FYM

Partially rotten farmyard manure has to be applied three to four weeks before sowing while well-rotten manure can be applied immediately before sowing. The existing practice of leaving manure in small heaps scattered in the field for a very long period leads to loss of nutrients. These losses can be reduced by spreading the manure and incorporating by ploughing immediately after application. The entire amount of nutrients present in farmyard manure is not available immediately. About 30 per cent of nitrogen, 60 to 70 per cent of phosphorous and 70 per cent of potassium are available to the first crop.

Compost

A mass of rotted organic matter made from waste is called compost. The compost made from farm waste like sugarcane trash, paddy straw, weeds and other plants and other waste is called farm compost. The average nutrient contents of farm compost are 0.5 per cent N, 0.15 per cent P_2O_5 and 0.5 per cent K_2O. The nutrient value of farm compost can be increased by application of superphosphate or rock phosphate at 10 to 15 kg/t of raw material at the initial stage of filling the compost pit.

Phospho-compost

Phospho-compost or P-enriched compost can be made by mixing rock phosphate @ 5% P_2O_5 with the composting mass. The addition of insoluble source of P to enrich compost is a more rational and practically useful approach since solubilization of insoluble P occurs during composting. Following compostable materials consisting of crop residues, grasses, weeds, tree leaves, animals feed wastes or their mixtures and cattle dung are used along with P sources.

Green manure

Green manuring can be defined as a practice of ploughing or turning into the soil un-decomposed fresh green plant tissue for the purpose of improving fertility status and physical structure of the soil. The increase in yield after green manuring is usually 30-50%. The fertilizing value of the legume crops can be increased greatly by manuring it with super phosphate. This practice not only increases the P content of the green manure but also encourages their plant growth on the whole, thus converting inorganic fertilizer into organic manure. A leguminous crop of 45 to 60 days can add upto more than 200 kg/ha when ploughed under; generally it is around 100 kg/ha green manure N, which corresponding to the average amount of mineral fertilizer N applied to most agricultural crops, is as efficient as fertilizers N.

In-situ Green manuring: Green manuring crops are grown and buried in the same field which is to be green manured, either as a pure crop or as an inter-crop with the main crop. The most common green manure crops in this system are sunnhemp (*Crotolaria juncea*), dhaincha (*Sesbania aculeata, Sesbania rostrata*) and cluster bean (*Cymopsis tetragonoloba*). *Dhaincha*, sunhemp and cowpea are the major green manure crops capable of accumulation of 4-5 tons/ha of dry biomass and 100 N kg/ha in 50-60 days, when well-managed. In Sikkim's conditions, rice bean and buck wheat can be recommended as green manuring crops.

Ex-situ Green Leaf Manuring: It refers to the collection leaves and tender twigs from shrubs and trees grow on bunds, wasteland and nearby forest areas and then incorporate them into cultivable field. The common shrubs and trees used for introduced green leaf manuring are *Gliricidia sepium, Sesbania speciosa,* karanj (*Pongamia glabra*), *Ipomea, Jatropha gossipifolia.*

Sheep and goat manure

The droppings of sheep and goats contain higher nutrients than farmyard manure and compost. On an average, the manure contains 3 per cent N, 1 per cent P_2O_5 and 2 per cent K_2O. It is applied to the field in two ways. The sweeping of sheep or goat sheds are placed in pits for decomposition and it is applied later to the field. The nutrients present in the urine are wasted in this method. The second method is sheep penning, where in sheep and goats are kept overnight in the field and urine and fecal matter added to the soil is incorporated to a shallow depth by working blade harrow or cultivator.

Poultry manure

Poultry manure contains higher nitrogen and phosphorus compared to other bulky organic manures. The average nutrient content is 3.02% N_2; 2.63 per cent P_2O_5 and 1.4% K_2O. Except for nitrogen, the availability of most nutrients in poultry manures is fairly consistent. Thus, soil mixing can reduce nitrogen losses and increase the amount available to plants.

Concentrated organic manures

Concentrated organic manures have higher nutrient content than bulky organic m anure. The important concentrated organic manures are oilcakes, blood meal and fish manure. These are also known as organic nitrogen fertilizer. These organic fertilizers are, therefore, relatively slow acting, but they supply available nitrogen for a longer period.

Oil cakes

After oil is extracted from oilseeds, the remaining solid portion is dried as cake which can be used as manure. The oilcakes are of two types: *Edible oilcakes* which can be safely fed to livestock; *e.g.*, Groundnut cake, Coconut cake, etc., and *non-edible oil cakes* which are not fit for feeding livestock; *e.g.*, Castor cake, Neem cake, Mahua cake. Both edible and non-edible oil cakes can be used as manures. However, edible oil cakes are feed to the cattle and non-edible oil cakes are used as manures especially for horticultural crops.

Bio-fertilizers

Bio-fertilizers (BF) is the product containing cells of different types of micro-organisms that have an ability to mobilize nutritionally important elements from non-usable to usable form through biological process. Although the potentiality of bio-fertilizers has increased in India in the recent years, there exists a large gap between its demand and supply. Since it is not a plant food it requires careful management and nutrients like other organisms for proper functioning. Their success in fixing atmospheric N and solubilizing insoluble P into plant utilizable P in field conditions depend mainly on the development of raw material as carrier which is the most important ingredient in bio-fertilizer. Some bio-fertilizers are *Azolla*, blue-Green algae (BGA), *Rhiziobium, Azotobacter, Azospirillum* and others. Generally, 200 g bio-fertilizer is suspended in 200 to 400 ml water to make a slurry, which is poured on 1-10 kg seeds slowly and mixed thoroughly with culture. Inoculation of seeds with efficient strains of *Azospirillum* and *Azotobacter* contributes about 15-20 kg N/ha in different crops.

Phosphate solubilizing micro-organism

Larger percentage of soil phosphorus is in the unavailable form and not more than 1-2 per cent, if it is incorporated into the above ground parts of the plants. Phosphorus solubilizing micro-organism includes various bacterial, fungal and actinomycetes forms that help convert insoluble inorganic phosphate into simple and soluble forms. Members of *Pseudomonas, Micrococcus, Bacillus, Flavobacterium, Penicillium, Fusarium, Sclerotium* and *Aspergillus* are some of the phosphate solubilizing micro-organisms. Integrated use of P solubilizing bacteria with rock phosphate contributes about 30-35 kg P_2O_5/ha.

Method of application of bio-fertilizers

Seed treatment for broadcasting: One packet of 200 g of bio-fertilizer is mixed in 500 ml of water to prepare slurry and the slurry is mixed with about 10-12 kg seeds and is air dried under shade. Soon after drying the bio-fertilizer coated seeds are broadcast.

Seedlings dipping for transplantation: In this method 1 kg bio-fertilizer that is sufficient for one acre is mixed in 5 lire water in a bucket and mixed thoroughly. The roots of seedlings bundles are dipped in the bio-fertilizer suspension for about 30 min. and these bio-fertilizer-coated seedlings are transplanted immediately.

Soil application: In soil application 4 to 5 kg of desired bio-fertilizer is mixed with 100 kg of well-decomposed FYM and mixed thoroughly to get a uniform mixture. The bio-fertilizer mixed FYM is applied to the soil 15 to 20 days of transplanting rice crops or one month after broadcasting of seeds of FYM is applied in the pits before transplantation or in dragged ring so as to ensure the placement near the root zone of the crop. Immediately after bio-fertilizer is applied, it should be covered with soil to avoid exposure to direct sunlight.

Placement of biofertilizers in Root Zone: Since the microbes like to grow and proliferate at the rhizosphere zone of the crop, the added inoculants should be placed close to the root zone so that the mutual requirement of microns and the beneficiary crop can meet and the inoculated microbe could easily established in the root rhizosphere.

Vesicular Arbascular Mycorrhizae (VAM)

The VAM is Vesicular Arbascular Mycorrhizae, called fungi which possess special structures known as vesicles and arbuscules. They help in the transformation of nutrients from soil to root system. They often increased uptake of nutrients and water. These fungi (VAM) are very suitable for groundnut, soyabean, millets, coffee, citrus, pepper and cloves. Mycorrhizae help in mobilizing insoluble soil phosphates. They further help increasing nutrient uptake. The fungal hyphae growing out from the plant root reach to the additional soil areas and help to absorb many nutrients for transmission to the plant, particularly less mobile nutrients, such as phosphate (PO_4^{3-}), Zn^{+2}, Cu^{+2} and MoO_4^{-2}.

Nutrient recycling

It is evident that application of FYM, green gram mulching, *Glyricidia* green foliage loppings and sunnhemp as GM increase yield over control and considerable build up of soil available nutrients following these. Besides, direct addition of N to the available soil pool, organics facilitates in the greater multiplication of soil microbes that could convert organically bound N to inorganic form there by help maintaining/restoring soil N status. Similarly, microbial decomposition of organics could form organic complexes with sesquioxides and thus, reducing the P fixing capacity of this soil. Moreover, enhancement in K availability is fortified following mineralization of organics and release of K to soil pool due to organic matter-clay interaction and reduction in K fixation.

Vermicompost

Compost that is prepared with the help of earthworm (may be local species or more vigorous exotic ones) is called vermicompost. Vermicompost is popularly known as *Black Gold* due to its colour and quality in supplying nutrients to plants, enriching soils by improving physicochemical and biological properties. Earthworm consumes large quantities of organic matter and excretes soil as casts. Casts of the earthworm have several enzymes and are rich in plant nutrients, beneficial bacteria and mycorrhizae. The nutrients value in casts depends upon the source of raw materials and the species of earthworm used. A fine worm cast is generally rich in N, P, K other nutrients in trace amount. Nutrient analysis of bio-fertilizer enriched vermicompost made from *Eisenia foetida* is pH 6.8, OC 11.9%, OM 20.5%, N 2.0%, P 1.0% and K 2.0%. The rate of application of vermicompost depends upon the nature and requirement of crops. General recommended dose is for field crops (6-8 t/ha), fruit crops (3-5 kg/ha) and pots (100-200 g/pot).

Soil acidity management

The main objective of acid soil management is to maintain appropriate soil pH in order to assure multi-condition favourable to plants. The desirable soil pH range for most of the crop is 6.0-7.0. The management of acid soils can be solved either by addition of lime/dolomite to counteract the abnormality of acid soils or to manipulate the agricultural practices under the stress of soil acidity. Grasses tend to tolerate acidic soils better than legumes. After soil testing, if the pH of the soil is less than 6.0 then only lime is recommended. Lime/

dolomite should be applied 10-15 days before sowing of crops either by broadcasting or by furrow application in 0-15 cm soil depth. During lime application, the greater the amount of moisture in soil are, more rapid is the rate of reaction. If the liming materials are finer, the effectiveness will be higher. The broadcast application of limestone @ 1-2 t limestone/ha gives the optimum yield in a sequence for two years and thereafter, half the amount of limestone is needed for sustained crop production. Furrow application of lime @ 250-400 kg/ha every year is economical. The poor and marginal farmers with resource constraints can invest on 250 kg limestone/ha in furrows every year to achieve the optimum crop productivity. Furrow application is very effective and requires lesser amount. Application of organic matter improves buffer capacity of soil and reduces Al toxicity. However, higher yield can be achieved by the application of 1 t/ha limestone or dolomite with 10 t FYM/ha. Nitrogen fixation both by symbiotic and non-symbiotic is favoured by adequate liming. Club root disease of cole crops can be reduced with liming. But overliming results in deficiency of many nutrients. Wood ash can also be used to increase soil pH.

Biopesticides for insect-pest and disease management

Biopesticides are natural plant products that belong to the so-called secondary metabolites that include thousands of alkaloids, terpenoids, phenolics and minor secondary chemicals. These substances have usually no known function in photosynthesis, growth or other basic aspects of plant physiology; however, their biological activity against insect-pets, nematodes, fungi and other organisms is well-documented. Every plant species has developed an inbuilt unique chemical complex structure that protects it from pests. The plant kingdom offers us a diverse array of complex chemical structures and almost every imaginable biological activity. These biopesticides are considered environmentally safe, selective, biodegradable, economical and renewable alternatives for use especially under organic farming systems.

Neem (*Azadirachta indica*) has emerged as the single most important source of insecticides made from isolates from different parts of the neem tree. These compounds and neem extracts exhibit almost every type of biological activity conceivable against a wide range of insects. More than 300 species of insects are reportedly affected by the neem components (Singh, 1993). Botanical insecticides are ecologically and environmentally safer and generally affect the behaviour and physiology of insects rather than killing them. Among the botanical pesticides investigated, neem has justifiably received the maximum attention during the last three decades. All parts of the neem tree possess insecticidal property but seed kernel is the most active. Neem bark, leaf, fruit and oil as well as extracts with various solvents especially ethanol have been found to exhibit activity against insect-pests. Neem product appear to be quite promising against such nefarious pests *Spodoptera litura, Helicoverpa armigera, S. gregaria* and others. Most of the stored grain pests have also been found to be responsive to neem products.

Crop losses due to insect pests and diseases are estimated to be Rs. 6000 crores annually in terms of agricultural crop produce. Though the crop productivity increased significantly due to use of mostly chemical pesticides, their continuous use contributed to increasing environmental threat to natural resources, wildlife, non-target beneficial species, the development of pest resistance/resurgence, etc. Mounting concern for environment has, therefore, led to evolve an alternate coherent pest management programme which is eco-friendly, self-sustaining and yet provide effective management of pests and diseases. In this context the use of biopesticides or pesticides of microbial origin are becoming increasingly important. These biological products are eco-friendly, possess higher degree of host specificity, genetically stable and do not pose the risk of resistance development. They are compatible with bio-fertilizers and many agro-chemicals, and can be exploited under the ambit of Integrated Pest Management (IPM). At present, more than 21,500 naturally occurring microbes or their metabolites are known to possess pesticidal properties. There is an increasing trend to minimize the use of pesticides. The world agro-chemical market is currently estimated to be $26,800 million, whereas bio-pesticide sales are around $120 million. Bio-pesticides sales, however, are estimated to increase at 10-25% pre annum in contrast to static or shrinking world agro-chemical market.

Perhaps it is high time to view agriculture without chemicals/synthetic pesticides. It can only be possible by discovering/formulating highly active bio-pesticides and popularizing them among farmers for their use. Undoubtedly, bio-pesticides will be the focus of main attention for plant protection practices to save agro-environment and humanity in 21st century. Due to higher potential of microbes in sustainable agriculture, they are currently being developed at large scale in the laboratory and pilot production units for commercial use in field.

Viral bio-pesticides

Viral bio-pesticides of baculovirus group, namely nuclear polyhedrosis viruses (NPVs) and granulosis viruses (GVs) offer great scope as crop protection agents on high value crops, such as cotton and vegetables against lepidopteran pests like *Helicoverpa armigera* Hbn. and *Spodoptera litura* Fabr. More than 500 baculoviruses (BVs) have been reported so far. Some of the insect pests from which occlusion viruses have been reported in India, have great potential in IPM.

Nuclear polyhedrosis virus

Nuclear polyhedrosis viruses kill the insets, when they are ingested. The viruses rapidly multiply and spread into cell nuclei. Diseased larvae move frequently to the periphery of the plant and die showing symptoms by hanging from the plant upside down in the form of 'V' called *Wipfelkrankit* (top disease) by German workers. The mass multiplication technique of virus has been well-demonstrated using 4th instars larvae of *Helicoverpa* and *Spodoptera.* Fully infected larvae shield on an average 2x10 poly occlusion bodies (POB) per larva. Two well-timed application of NPV @ 1.5x10 POB (250LE ha) in the evening hours (to avoid ultraviolet inactivation) coinciding with early stages of the larvae successfully control the pests on chickpea, pigeon pea and sunflower. Similarly, this could be managed on several crop hosts by applying NPV@ 250-500 LE ha depending upon crop canopy.

In India, NPVs of *H. armigera* and *S. litura* have been formulated as basic formulations (talc based) like dust, wetable powder and ultra-low volume (ULV) product and have been found effective against these pests on several crops. ULV application of virus in 10% whole egg homogenate @250LE ha have proved effective against *Helicoverpa* on pigeon pea Jayaraj *et al.* (1992). Eclar is the first commercial viral insecticide based on NPV developed in 1961. Recently some commercial liquid formulations like H-NPV, Helivax; Biovirus, Heloionil and S-NPV have appeared in the market for biocontrol of notorious pests like *Helicoverpa* and *Spodoptera.*

Granulosis virus

Next important to NPVs is granulosis viruses. The GV of sugarcane internode borer has been extensively studied in India. A spray of 10^9

occlusion bodies (Obs) ml^{-1} or combination of 10^5 OB ml^{-1} + 0.05% endosulfan effectively controls the pest. It is reported to be safe to natural enemies occurring in sugarcane ecosystem.

Bacterial bio-pesticides

Many spore former and non-spore forming bacteria are known to be effective against a wide spectrum of insects and pests. The crystalliferous *Bacillus thuringiensis* (Bt) has been found to be effective against several species of lepidopterous pests. Its insecticidal activity is primarily caused by parasporal crystal (delta endotoxin) produced during sporulation. The discovery of isolate capable of acting against coleopterans (*B. thuringiensis tenebrionis* and *B. sphaericus* against mosquito and *B. moritai* against dipteran insects has further extended the scope of bacterial insecticides. Bt (300g $ha^{-1)}$ has further been used in the management of several pests notably diamond black moth *Plutella xylostella* on cruciferous vegetables. Certain adjuvant, like whole egg homogenate 1% and whole milk (0.5%) added to Bt, increase the efficiency. It is compatible with several insecticides, extract of neem seed jerbel (5%) and *Catherebthus roseus.* Bt @ 0.5-1.0kg formulation ha^{-1} has been found quite effective against important pests like *S. litura, E.vitella, H. armigera* and *P. xylostella.* The work on *B. Popillae* is limited. It causes disease in white grubs. It is reported from several parts of the country and dispersed by birds feeding on grubs and also through irrigation water. If established, it maintains itself in the area for many years.

Several bacteria, like, *Pseudomonas fluorescens, Bacillus subtilis, Streptomyces nigrifaciens, Agrobacterium radiobacter and Azotobacter* sp. are known to be potential bioagents against several plant pathogens. Most of these biocontrol bacteria are, however, in the stage of empirical application, either in the green house or in the field with a few exceptions. There are reports of bacterial bioagents in enhancing the growth and greater root volumes of crop plants, when applied as seed treatment or soil application. This envisages that the responses may be as bio-fertilizer or by the control of some undiagnosed plant pathogens. Recently, *Pseudomonas* sp. is being promoted as plant growth promoting rhizobacteria (PGPR). The mechanism of pathogen suppression by these bacteria includes substrate competition, niche exclusion, and production of siderophores, antibiosis and induced resistance (Rai, 2003).

Fungal bio-pesticides

Fungi unlike bacteria or virus do not require ingestion for infection; so sucking pests are also targeted by primary contact or by secondary uptake from sprayed vegetation. The pathogenesis begins with germination of conidia on the cuticle and penetration and development inside the host leading to death of host essentially under high humid condition.

Of the several fungi infecting insects, the green muscardine fungus, *Metarhizium anisopliae* has shown promise against several insect pests. *M. anisopliae major* @ 5x10/mu[23] effectively controls rhinoceros beetle in manure pits. Whereas, *M. anisopliae* has been found very effective against BPH, WBH and leaf folder on rice. *Fusarium oxysporum* @ 6.25×10^6 spores ml^{-1} caused 100%wp (*Pandora delphacis* based) applied at 30 and 50 DAT effectively checks BPH population (87.8% mortality). The white muscardine fungus, *Beauvaria bassiana* has also great potential in the control of rice pests like, BPH, WBPH and rice ear head bug as well as sweet potato weevil *Cylas formicarlus*. A few myco-insecticides have recently come in market to thwart the pest problem.

Biological control of plant diseases is mainly limited to soil borne diseases. However, studies on seed borne and air borne diseases are underway in India. The successful use of biocontrol fungi *Trichoderma Viride, T. harzianum. T. virens and Trichoderma* sp are well-documented against soil borne disease of various crops like rice, groundnut, pea, grams, ginger, horticultural crops and vegetables in India and abroad. Recently, some other fungi like, *Aspergillus niger, A terreus, A. nidulans* and vesicular-arbuscular mycorrhizae (VAM) have also shown promise against soil borne diseases. Broad-spectrum activity of *A. niger* AN-27 against *Pythium, Fusarium, Macrophomina, Rhizoctonia* and *Sclerotinia* is also suggesting their use as potential biocontrol agents (Rai, 2003).

Why Bio-pesticides?

- Human and environmental safety
- Alternatives to conventional pesticides
- Amenable to small-scale, local production in developing countries
- Address increased public awareness of environmental and food safety

- Fundamental component of Integrated Pest Management
- Products available in small, niche markets that are typically un-addressed by large agrochemical companies

Bio-pesticides are pest management tools that are based on beneficial micro-organisms (bacteria, viruses, fungi and protozoa), beneficial nematodes or other safe, biologically based active ingredients. Benefits of bio-pesticides include effective control of insects, plant diseases and weeds, as well as human and environmental safety. Bio-pesticides also play an important role in providing pest management tools in areas where pesticide resistance, niche markets, and environmental concerns limit the use of chemical pesticide products.

Types of control	Examples
Insect Control Bacteria	*Bacillus thuringiensis, B. sphaericus, Paenibacillus popilliae, Serratia entomophila*
Viruses	Nuclear polyhedrosis viruses, granulosis viruses, non-occluded baculoviruses
Fungi	*Beauveria* spp, *Metarhizium, Entomophaga, Zoopthora, Paecilomyces fumosoroseus, Normuraea, Lecanicillium lecanii*
Protozoa	*Nosema, Thelohania, Vairimorpha*
Entomopathogenic nematodes	*Steinernema* spp. *Heterorhabditis* spp.
Others (not strictly bio-pesticides)	Pheromones, parasitoids, predators, microbial by-products
Weed Control Fungi Bacteria	*Colletotrichum gloeosporioides, Chondrostereum purpureum, Cylindrobasidium laeve* *Xanthomonas campestris pv. poannua*
Plant Disease Control Fungi Competitive inoculants Composts, soil inoculants	*Ampelomyces quisqualis, Candida spp. Clonostachys rosea f. catenulate, Coniothyrium minitans, Pseudozyma flocculosa, Trichoderma* spp *Bacillus pumilus, B. subtilis, Pseudomonas* spp, *Streptomyces griseoviridis, Burkholderia cepacia*
Nematicides, ***etc.***	
Nematode trapping fungi	*Myrothecium verrucaria, Paecilomyces lilacinus*
Bacteria	*Bacillus firmus, Pasteuria penetrans*
Mollusc parasitic nematode	*Phasmarhabditis hermaphrodita*

Promotion of Organic Farming

The Government of India is extremely concerned about the soil health status of India vis-à-vis production of quality food. Intensive cropping, soil erosion, deforestation, pollution have the potential to reduce soil health that can influence the food productivity. Keeping this in view, GoI initiated the Soil Health Scheme and promoting organic farming in rainfed and hilly areas, where fertilizer consumption is low through various schemes/programmes under National Mission for Sustainable Agriculture (NMSA)/Paramapragat Krishi Vikas Yojana (PKVY), Rashtriya Krishi Vikas Yojana (RKVY), Mission for Integrated Development of Horticulture (MIDH), National Mission on Oilseeds & Oil Palm (NMOOP) and Network Project on Organic Farming and Network Project on Organic Horticulture of ICAR. Government is implementing a Cluster based programme to encourage the farmer for promoting organic farming under PKVY. In order to implement the PKVY in the year 2015-16, an amount of Rs.300 crore has been allocated with the following objectives:

- Groups of farmers would be motivated to take up organic farming under Paramparagat Krishi Vikas Yojana (PKVY). Fifty or more farmers will form a cluster having 50 acre land to take up the organic farming under the scheme.
- In the next three years (2015-16 to 2017-18) the government is targeting formation of 10,000 clusters covering five lakh acre area under organic farming. There will be no liability on the farmers for expenditure on certification.
- The cluster chosen for Organic Farming shall be to the extent of 50 acres.
- Under the scheme every farmer can get Rs. 20,000 per acre in three years for seed to harvesting of crops and to transport produce to the market.
- Further, to make the scheme work end-to-end, government will promote Organic Farming using traditional resources, adopting eco-friendly low cost technologies and for higher returns, the organic products will be linked with the market.
- It will increase domestic production and certification of organic producc by involving farmers.

National Project on Organic Farming (NPOF) is a continuing central sector scheme since 10th Five Year Plan. NPOF is being implemented by National Centre of Organic Farming at Ghaziabad and its six Regional Centres at Bangalore, Bhubaneshwar, Panchkula, Imphal, Jabalpur and Nagpur. For carrying out instensive research on organic production system and further promotion of organic farming in the country, the ICAR Research Complex for NEH Region, Sikkim Centre has been upgraded into National Organic Farming Research Institute on 16 May, 2016. All the institutes are promoting organic farming in the country with the following major objectives:

- Maintenance of soil fertility by encouraging and enhancing the biological cycle within farming systems involving micro-organisms, soil flora and fauna, plants and animals.
- Identification of areas and crops suitable for organic farming.
- Development of organic package of practices.
- Setting up of model organic farms for getting seed material for organic cultivation.
- Assurance of production and supply of quality organic input.
- Adoption of biological methods for pest and disease control.
- Adoption of biological and mechanical methods for weed management.
- Harnessing of traditional and indigenous knowledge relating to organic farming.
- Creation of awareness among farmers towards organic agriculture.
- Development of domestic market for organic produce.
- Improvement of farmers' income through production of quality produces.
- Generation of rural employment opportunity.
- Simplification of certification system and recognition of adequate certification agencies, especially for domestic market.
- Promotion of Organic Farming through low cost certification system known as "Participatory Guarantee System".
- Maintaining a diversity of plant and animal species as a basis for ecological balance and economic stability.

- Improvement in condition of livestock that allow them to perform all aspects of their innate behaviour.
- Development of regulatory mechanism for various organic input and organic produce.
- Promotion of organic farming in the country through technical capacity building of all the stakeholders including human resource development, transfer of technology, promotion and production of quality organic and biological inputs.
- Awareness creation and publicity through print and electronic media.
- Revision of standards and testing protocols keeping in view the advances in research and technology and bringing remaining organic inputs under quality control regime.
- Organic input resource management, technology development through support to research and market development.

Organic Standards and Certification

India has evolved the national standards for organic products keeping in view the traditional with the scientific approach. These standards are in conformity to the international standards of the European Union (EU) and International Federation of Organic Agriculture Movements (IFOAM).

Government of India has identified six Accreditation Agencies for certification or organic products. They are APEDA (Agricultural and Processed Food Products Export Development Authority, under the Ministry of Commerce and Industries, GoI), Tea Board, Coffee Board, Spices Board, Coconut Development Board and Directorate of Cashew and Cocoa Board under the Ministry of Commerce and Industries. These Accreditation Agencies will strive to achieve a socially beneficial goal through the certification programmes combined with a collective quality symbol. Based on the criteria drawn for accreditation, the Accreditation Agencies in turn would identify the inspection and certifying agencies for accreditation of their certification programme. There are 28 certifying bodies accredited by APEDA. The list is available on the APEDA website under National Programme on Organic Production (NPOP). The costs and tariffs are also available on the website for interested parties. They will operate certification programmes in the following areas:

1. Organic production
2. Organic processing
3. Providing test standards

Inspections will be carried out by the accredited inspection and certifying agencies at all stages of production and processing with regular check on all operators (farmers and processors) in accordance with the principles in the basic standards.

Organic products produced under the national standards would be certified and approved by the accredited inspection and certifying agencies for labeling as "India Organic". The labelled products would provide the consumer with the best possible guarantee of organic origin, production, processing and packaging (Gauri, 2004). The basic standards are provided in the pages to follow. The following foreign certification agencies have been accredited for certification by APEDA. Detailed organic production standards are available in the NPOP Guidelines on APEDA website as VII edition, 2014 (www/apeda.gov.in/).

Constraints of Organic Farming

Lack of awareness

- Lack of skill about improved methods of compost making
- Lack of awareness about the concentration, time and method of manures and bio-fertilizer application
- Lack of proper training about organic farming
- Inadequate knowledge of field functionaries about organic farming
- Lack of knowledge in farmers about different bio-pesticides
- Farmers think that chemical fertilizers are more effective than manures and bio-fertilizers
- Low credibility of source for purchasing compost and bio-fertilizers

Marketing problems: It is found that before the beginning of the cultivation of organic crops, their marketability and that too at a premium over the conventional produce has to be assured. Inability to obtain a premium price, at least during the period required to achieve the productivity levels of the conventional crop will be a setback.

Shortage of biomass: The small and marginal cultivators have difficulties in getting the organic manures compared to the chemical fertilizers, which can be bought easily, of course if they have the financial ability. But they have to either produce the organic manures by utilizing the bio-mass they have or they have to be collected from the locality with a minimum effort and cost. Increasing pressure of population and the disappearance of the common lands including the wastes and government lands make the task difficult. Besides, very high volume of organic materials required to meet the nutrient demand. [*e.g.*, Average application of 234 kg N/ha/year for a system, this requires either 46 t of raw FYM (0.5 % N) or 15.5 t vermicompost (1.5 % N)].

Inadequate supporting infrastructure: The certifying agencies are inadequate, the recognized green markets are non-existent, the trade channels are yet to be formed and the infrastructure facilities for verification leading to certification of the farms are inadequate.

Absence of an appropriate agriculture policy: Promotion of organic agriculture both for export and domestic consumption, the requirements of food security for millions of the poor, national self-sufficiency in food production, product and input supplies, etc. are vital issues, which will have to be dealt with in an appropriate agriculture policy of India.

Inability to meet the export demand: The market survey done by the International Trade Centre (ITC) during 2000 indicates that the demand for organic products is growing rapidly in many of the world markets while the supply is unable to match it. India is known in the world organic market as a tea supplier and there is a good potential to export coffee, vegetables, sugar, herbs, spices and vanilla. In spite of the several initiatives to produce and export organic produces from the country, the aggregate production for export came to only about 14000 tons. This also includes the production of organic spices in about 1000 ha under certification. The country could export almost 85 per cent of the production indicating that demand is not a constraint in the international markets for organic products.

Risk Associated with Organic Farming

Some of the risks associated with organic farming are as follows:

1. Risks of contamination of organic crops by gentically modified organisms (GMOs);

2. Shortages of particular inputs such as certified organic seeds and biological pesticides;
3. Access to capital, because banks are sometimes unfamiliar with organic production systems;
4. Instability of organic price premiums; and
5. Some crops in organic rotations do not benefit from USDA commodity program price and income protection.

Farmers' Experience

Promotion of Integrated organic farming system for sustaining income, food and nutritional security of small and marginal farmers.

IOFS model (<0.50 ha)

Integrated organic farming system model for constant income was developed (0.43 ha) at the Agronomy Research farm of ICAR Research Complex for North-Eastern Hill (NEH) Region, Umiam, Meghalaya. The IOFS model comprising different enterprises such as cereals (rice and maize), pulses (lentil, pea), oilseeds (soybean, rapeseed), vegetable crops (Frenchbean, tomato, carrot, okra, brinjal, cabbage, potato, broccoli, cauliflower, chili, coriander, etc.), fruits (Assam lemon, papaya, peach), dairy unit (a milch cow + calf), fodder crops, central farm pond, farmyard manure pits and vermicomposting unit. A farm pond of 460 m^2 area with average depth of 1.5 m was part of the IOFS model for life saving irrigation and aquaculture. Climbing vegetables such as bottle gourd, chow-chow, cucumber, ridge gourd, etc., were grown on a structure created above water bodies in one side of the pond dyke for vertical intensification. Pumpkin was raised in another side of the pond and allowed to crawl on the ground. The washings from the dairy unit were diverted to fish pond for promoting growth of zooplankton and phytoplankton for fish growth. The solid waste from cow shed was used for FYM making and vermicomposting.

The total cost of cultivation was recorded at Rs. 55,839/- per year under the IOFS model with an area of 0.43 ha. Maximum expenditure was incurred in crop component of the model with 48% of the total cost of cultivation. Dairy unit with one adult cow and one calf registered 36 % of the total cost of cultivation, while fishery component recorded 9 % of the total cost of cultivation (Table 27). For maintaining vermicomposting unit of 72 m^2 area and other important operations like hedgerow planting, residue recycling, rock phosphate application and liming, the expenditure incurred was Rs. 3700/- which account to 7 % of the total cost. Since, the soil of the north-east India and study site is acidic in nature, application of rock phosphate (100-150 kg/ha) and liming (500 kg/ha in furrow in alternate years) is required for phosphorus nutrition as application of organic manure alone is not enough to meet the crop requirement. A total net return of Rs. 62,531/- per year was achieved under the IOFS model which is much higher than the region's farmer common practices of rice monocropping or improved practice of rice-vegetables cropping system. The highest contribution towards the total net return was contributed by crop component of the model (61%), followed by dairy (25%) and fishery component (20%). The fish production was 136 kg. The net return from dairy component was calculated only in terms of milk production since the cow dung produced, was recycled back into the model which was used as manure for crop production. The production of vermicompost from model was 1500 kg annually and it was used in the farm itself for nutrient supplement to crops.

Organic Food Production Through Integrated Farming System-Cluster Approach (case study)

A village (Mynsain) under Ribhoi District and Umsning block in Meghalaya (25^{0}44.339'N, 092^{0}00.937 E and 884 m) block have been adopted for disseminating organic production technology in a cluster approach. The village is 20 km away from the institute (ICAR RC for NEH Region, Umiam), having 120 households with an approximate area of 60 ha. As per the interaction with the farmers and elderly peoples of the village, it is learnt that the village is totally organic and so far no inorganic input has been applied. The sensitization meeting with the villagers including village head (Headman), member of the SHGs, Department of agriculture were organised to create awareness on organic farming. The improved seeds and planting materials (Maize, Groundnut, Frenchbean, tomato, carrot, etc.) were distributed to the farmers. A formal Memorandum of Understanding (MOU) between ICAR and the village was made. The survey (PRA) and farmers training were conducted to initiate the programme. The model has following components:

- Organic food production: Cereals, pulses, oilseeds, vegetables and fruits

- Food-Feed Crop Production: food for consumption purpose and as feed for livestock.
- Livestock: Piggery and dairying. Backyard poultry farming.
- Community vermicomposting unit: Biomass recycling and quality manure production
- Green manuring and green leaf manuring: Generation of organic manure in system.
- Hedge row intercropping: *Tephrosia, Crotolaria, Indigofera* spp. as fencing, conserve soils and water and supply nutrient rich green leaf manure.
- Planting of multipurpose tree species (MPTS), bamboos, etc.: conserving soil, generating additional income as well as for environmental security.
- Development of water harvesting structure: Farm ponds and jalkunds.
- Soil conservation measures: Terracing, half-moon terracing, vegetative barriers, etc. Cultivation of fodder crops in degraded lands: To supply green and rehabilitation degraded land.
- Organic outlet: Near highway for marketing organic produce.
- Capacity building: Training and field visits were organized to enhance the capacity of the farmers in organic farming technologies.

Integrated Organic Farming System Improves Income and Livelihood Security in Sikkim-A success story (http://www.icar.org.in/en/node/10912 and http://www.kiran.nic.in/OFS_sikkim.html)

Shri Nim Tshering Lepcha, resident of Lower Nandok under 26 Naitam-Nandok GPU, East Sikkim had been practicing traditional agriculture in his 2 ha land as the only means of livelihood. Despite hard labour, the farm productivity was low and income was not satisfactory. But the interventions during last three years (2013-16) by Krishi Vigyan Kendra, ICAR Sikkim Centre, Ranipool, East Sikkim have transformed his socio-economic status.

Technological Interventions

Organic farming technological backup by ICAR Research Complex for NEH Region, Sikkim Centre made developmental interventions by providing training, on field demonstrations and input support. Various inputs/interventions were provided under National Innovations on Climate Resilient Agriculture (NICRA) with the purpose of re-orienting his traditional farming into integrated organic farming system (IOFS) to increase the farm income. Major among these are Agri-polythene sheets (250 GSM) for the purpose of making *Jalkund,* micro-rainwater harvesting structure; low cost plastic tunnels (transparent UV stabilized sheet of 45 GSM) for sequential vegetable cultivation, garden pea (TSX-10) under zero-till in rice-fallow rotation, cultivation of improved maize line RCM 1-1 and improved rice line RCM-10, backyard poultry production with Vanara ja and Hybrid Napier cultivation as fodder grass on terrace risers. Scientific management practices of fisheries with Grass carp and Common carp, milk cow crossbred Jersey and large cardamom cultivars *Sawney* and *Varlangey* were also introduced. Vegetables *viz.,* cabbage var. Rare Ball, cauliflower var. Suhasini, broccoli var. Everest, tomato var. Arka Samrat, coriander var. Super Midori, spinach var. All Green, radish var. Chinese White were sequentially cultivated under low cost plastic tunnels. *Jalkund,* a lifesaving water reservoir designed with dimens ions of 5 m x 4 m x 1.5 m (capacity of 30,000 l.) proved to be an indispensable tool in meeting the water requirement of crops through gravitational sprinkler irrigation system and encouraged the farmer to opt for diversification of the integrated organic farming system.

Impact

Capacity building of the farmer coupled with his own competence, enthusiasm & institutional intervention raised his stature as progressive farmer in Integrated Organic Farming System within short span of time. Presently, Mr. Lepcha earns gross income of Rs. 4,15,050/- [(rice cultivation (0.25 ha) = Rs. 8,000; maize cultivation as green cob and grains (0.25 ha) = Rs. 9,500; sequential vegetable production (seasons/year) under plastic tunnel (0.05 ha) = Rs. 1,50,000; Rabi vegetables (zero till cultivation of garden pea and cole crops in 0.15 ha) = Rs. 12,500; large cardamom production (0.25 ha) = Rs. 45,000; milk production (2,880 l/year) = Rs. 75,800; backyard poultry (Vanaraja 100 nos.) = Rs. 63,750; fisheries (500 nos. Grass carp and

Common carp) = Rs. 50,500/-] as compared to Rs.1,65,000/- under traditional farming system. An expenditure of Rs. 89,750/- was incurred (excluding farming family energy input i.e., labour cost) for maintaining the integrated organic farming system, resulting in benefit : cost ratio of 4.6.

He is now a role model not only in his village but also for the entire farming community of East Sikkim. Farmers from other villages are motivated by the success of this farmer and headmen of neighbouring villages are encouraging farmers to implement similar model in their villages. These interventions have the potential to create positive impact on the utilization of scarce resources under the fragile mountain ecosystem benefitting the farmers at large. In recognition of his achievements, Mr. Nim Tshering Lepcha was felicitated by Sh. Khorlo Bhutia, Secretary, Food Security and Agriculture Development Department, Government of Sikkim as progressive farmer on the 40th Foundation Day of ICAR Research Complex for NEH Region, Sikkim Centre, Tadong in 2015.

Conclusion

Organic farming offers the most sustainable solution for developing the agricultural sector and provides food security with least negative impacts on the environment and also offers solutions for sound rural development, besides provides healthy food, more ecosystem services and creating more jobs opportunities.

References

1. Chadha, K.L. *et al.*, 2001. Horticulture Development for X FYP. Report of Working Group Sr. No. 14/2001. Planning Commission, Govt. of India, New Delhi.
2. CAR, 2008. News reporter April-June 2008, pp 1-2
3. DRR, 2007, 2008. Progress report: Vol. III, Soil Science.
4. FAO, 2011. *Save and Grow - A policymaker's guide to the sustainable intensification of smallholder crop production*. Factsheets. Food and Agriculture Organisation of the United Nations, Rome.
5. FiBL and IFOAM, 2016. Organic International: The World of Organic Agriculture. Statistics and Emerging Trends. 340p. (www.organic-world.net/yearbook-2016.html)

6. Gay, S., Louwagie, G., Sammeth, F., Ratinger, T., Marechal, B., Prosperi, P., Rusco, E., Terres, J., Van Der Velde, M., Baldock, D., Bowyer, C., Cooper, T., Fenn, I., Hagemann, N., Prager, K., Heyn, N. and Schuler, J. 2009. *Final report on the project 'Sustainable Agriculture and Soil Conservation' (SoCo)*. EUR - Scientific and Technical Reports JRC51775, European Commission, Luxembourg.

7. Gouri, P. V. S. M. 2004. *National Programme for Organic Production*. *In* Anil Kumar Singh, Narayanasamy, G., Rattan, R. K. and Goswami, N. N. (*Eds*) *Organic Farming*. ISSS Bulletin No. 22. Indian Society of Soil Science, IARI, Pusa, New Delhi-110 012. 107p.

8. Handbook of Agriculture published by Indian Council of Agricultural research.

9. IPCC, 2007. Chapter 8: Agriculture, in *Climate Change 2007: Mitigation of Climate Change,* Working Group III contribution to the IPCC Fourth Assessment Report ed. Inter-governmental Panel on Climate Change, Geneva.

10. Integrated Nutrient Management through ICT by D.S. Yadav in GNFC, Bharuch

11. Kim, K.H. 2005. Two Concepts of Sustainability: Evaluation of Organic Farming. Deptt. of Agricultural and Resource Economics, Graduate School of Agriculture and Life Sciences, The University of Tokyo.

12. Meyer, R., Ratinger, T. and Voss-Fels, K.P. 2013. *Options for feeding 10 billion people - Plant breeding and innovative agriculture*. Report prepared for STOA, the European Parliament Science and Technology Options Assessment Panel, under contract IP/A/STOA/FWC/2008-096/LOT3/C1/SC3, Institute for Technology Assessment and System Analysis (ITAS), Karlsruhe Institute of Technology, member of ETAG, the European Technology Assessment Group, Available at: [http://www.europarl.europa.eu/stoa/cms/studies].

13. Poll, R.V.D and M. Smith. 2009. Basic Principles for Organic Cultivation and Livestock Production. Manual for Small Scale Farmers to start Organic and Livestock Production, Surplus People Project, Athlone, South Africa.

14. Principles of Agronomy by T.Y. Reddy and G.H.S. Reddi by Kalyani Publishers.

15. Rai, A.B. 2003. Bio-pesticides: an important component of integrated pest and disease management in sustainable agriculture. *In* A. K. Yadav (*Ed*) Organic Farming for Sustainable Agriculture. Regional

Bio-fertiliser Development Centre, State Agriculture Farm, Imphal, Manipur, India, pp. 28-33.

16. Roy, R.N., Finck, A., Blair, G.J. and Tandon, H.L.S. 2006. *Plant nutrition for food security - A guide for integrated nutrition management.* FAO Fertilizer and Plant Nutrition Bulletin 16, Food and Agriculture Organisation of the United Nations, Rome, Italy.

17. Singh, R.P. 1993. Botanical pesticides in developing countries: Current status and future trends. *In*: G. S. Dhaliwal and B. Singh. (*Eds*) Pesticides: Their Ecological Impact in Developing Countries. Commonwealth Publishers, New Delhi. pp. 236-269.

18. Sikkim AGRISNET. http://www.sikkimagrisnet.org/General/en/Default.aspx

19. Site-Specific Integrated Nutrient Management for Sustainable Rice Production and Growth by K.V. Rao in http://www.rkmp.co.in

❑❑❑

Chapter 7

Zabo: An Alternate to Shifting Cultivation for Livelihood and Resource Conservation in Nagaland

R.K. Singh, D. Borkotoky and P. Theluo

Krishi Vigyan Kendra, ICAR-NRC on Mithun, Porba, Phek, Pfutsero-797107, Nagaland

Corresponding author's mailing address, Senior Scientist and Head, KrishiVigyan Kendra Phek, ICAR-NRC on Mithun, Porba, Phek, Pfutsero-797107, Nagaland, Email: rksingh3@gmail.com

ABSTRACT

"Zabo" is an aboriginal way of water conservation and integrated farming system, which was evolved by the ingenious and skillful tribal farmers in Nagaland. The term 'Zabo' is derived from the word zabö, which is used for "impounding runoff water" in Chokeri dialect. It is also known as Dzüdü or Ruza system in certain areas of Phek district of Nagaland. Zabo system of farming has a combination of forestry, horticulture, agriculture, fishery and animal husbandry with well-founded soil and water conservation base. Considering the advantages of zabo system over the jhum, the present study was undertaken for systematically recording of various components practiced under zabo system and their significance in Kikruma village of Phek district, Nagaland. A total of 40 respondents were interviewed on various aspect of zabo farming system based on the questionnaire. The findings of the study revealed that cereals, pulses and vegetables are the major components of jhum, whereas in zabo, cereals, pulses, vegetable, fruits, livestock and fish are major component. Paddy is the major crop in zabo, however, maize and millets cover maximum acreage under jhum. The average yield of paddy in zabo system was recorded as 1.9 tons/ha against less than 1 ton/ha under jhum cultivation with higher seed requirements. The study showed that majority of farmers of the study area were having more satisfaction in practicing the zabo system than jhum due to the generation of better remuneration through animal husbandry, horticulture and fisheries. Zabo system is sustainable, requires less labour and more tolerant to climatic vulnerability. From the study, it can be concluded that zabo farming system is more remunerative, preserves soil fertility, eco-friendly as against to jhum system which leads to the wider destruction of area, soil erosion and creates a severe ecological imbalance as well as environmental pollution.

Keywords: Climatic vulnerability, Eco-friendly, Integrated Farming System, Jhum, Soil erosion, Water conservation and Zabo.

Introduction

Rice_(Oryza sativa L) is the main staple food of the people of North-east India and owing to varied physical conditions, such as topography, soil and climate there is so much diversity in the manner of rice cultivation. The region has uneven physiography, consisting both the valley and the hill so the people have different ways of cultivating the paddy as per the convenience and the resources available to them. Terrace and Jhum cultivation is the predominant form of cultivation in Nagaland. Jhum cultivation also called slash and burn or shifting cultivation is the older form of cultivation as compared to the terrace cultivation or wet rice cultivation. Whereas, zabo system is a comparatively recent intervention in terrace cultivation, which is prevalent in valleys dominated by Chakhesang and Angami tribes.

The term 'Zabo' is derived from the word 'zabö', used in Chokari dialect of Chakhesang tribe for "impounding runoff water". The practice was evolved by the people of Kikruma, which later spread to entire Chakhesang inhabited area of Phek, Nagaland. Zabo is also known as Dzüdü or Ruza in other parts of Chakhesang and Angami regions. It is a unique farming practice, which has a combination of forest, horticulture, agriculture, animal husbandry and fishery components with well-founded soil and water conservation base (Singh *et al.*, 2012). Water resource development, water management and protection of the environment are inherent aspects of the system (Sharma *et al.*, 1994). Though Kikruma area of the Phek receives enough rains but due to surface runoff region suffers from water scarcity that has forced the people to find an alternate way to mitigate, which has led to evolution of an elaborate and most efficient land-water management system "Zabo". It has an inbuilt water harvesting and recycling systems with well-founded conservation base to control soil erosion, management of soil fertility and available water. It is a viable practice of resource management and maintenance of ecological balance (Pulamte L., 2008).

Water harvesting ponds are constructed in the middle to collect the water from protected forest land. Hilltops above the pond, act as catchments area and water is channelized through inlet channels from the catchment area to the silt retention tanks, where it is stored for 2 to 3 days before transferring to the main ponds. This allows settling of the silt coming with runoff water. Silt retention tanks are cleaned annually and the soil dug from the silt retention tank is applied in the paddy fields, as it is rich in organic matter. Sometimes, more than one pond is constructed in such a way that the surplus water from one pond flows down to the other pond below. Fruit trees and other horticultural crops are planted at the bank or beside the ponds. Water is released from the pond for irrigation through an outlet at its base. The water from the ponds is transferred to fields either through open channels or by split bamboo channels. Animal enclosures fenced with ordinary woods and bamboo branches are constructed preferably below the water harvesting pond (Kithan, 2014). Water while passing through the animal yard water carries the dung and urine of the animals, thus it helps in maintaining soil fertility (Singh, A.K., 2007). Cross section of the entire continuum is elaborated as Fig.1 and photograph is presented as Fig. 2.

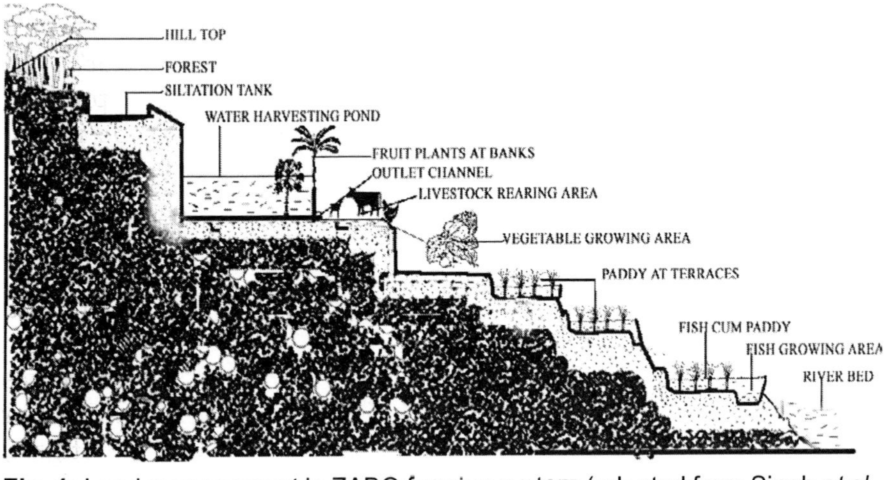

Fig. 1: Land management in ZABO farming system (adopted from Singh *et al.*, 2012)

Fig. 2: Photograph of a Zaboarea, showing catchment area, ponds, animal shelter and paddy field

Vegetables and fruits like squash, colocasia, cucurbits, banana, papaya, oranges and citrus are cultivated on the banks and besides the ponds. Vegetables are also cultivated just below the livestock enclosures (Fig. 3).

Fig. 3: Vegetables and banana cultivation on sides of the pond

Livestock like cattle, goat, sheep, pig and poultry are raised beside the pond (Singh *et al.*, 2012) (Fig. 4). Water stored in these ponds helps in providing an average of two supplementary irrigation to paddy fields in case of poor rains, which helps in getting yield between 3-4 tons per hectare (Sharma, U.C. and Sharma, V., 2003).

Fig. 4: Animal shelter in Zaboarea

Paddy cum fish culture is commonly practiced by the farmers in this region and farmers dug out a small pit in the middle of the rice field and fish fingerlings are released in the fields during the month of July. Paddy matures by the end of October and field dries by then or in case there is excess water, then it is drained out from the fields before harvesting of the paddy. As the field dries, fishes move into the pit and from there they are harvested (Fig. 5). On an average 50-60 kg of fish is harvested per hectare from paddy cum fish culture (Singh *et al.*, 2012).

Fig. 5: Fish harvesting from a paddy field in the Zabo area

Contrast to zabo, jhum cultivation involves the clearing of the jungle by slashing and burning down of trees, resulting in deforestation and serious environmental problem. Jhuming compromises soil stability, fertility and exposes it for degradation. Thus, people are becoming increasingly concerned and are looking for alternative means of efficient cultivation. Terrace cultivation is receiving wide appreciation and people are beginning to embrace terrace cultivation (Kithan, 2014). The practice of efficient farming system zabo, developed by the farmers of Kikruma is compared with jhuming widely practised in the north-eastern region.

Material and Methods

The study was conducted at Kikruma village of Phek district, Nagaland. The village was chosen purposely because of the belief of local people that the practice of zabo has started in this village and jhuming is still practised in some areas in and around Kikruma. The village is inhabited by Chakhesang tribe and they speak Chokri dialect. The questionnaire was prepared in English language and narrated to respondents in Chokri by the local youth from the same village. Most of the respondents were well-versed with the English language and in case any term was not clear to them, the same was explained to them in Chokri to make them understand better.

The data were collected through group discussions, structured interviews and questionnaires from 40 respondents of the village. Respondents were requested to tick the correct answer in the closed questions present in the questionnaire, however, for the open-ended questions, they were asked to answer in their own words. The study employed both primary and secondary data; both interviews and semi-structured questionnaires were used for collecting primary data on practices, yield and their perceptions about the practice of zabo. Secondary data was collected from the field through documentary analysis such as reports from the state government publications, journals, magazines, research reports as well as other publications.

Collected data were coded, entered in excel sheet, verified and cleaned before analysis. Microsoft Excel, 2013 Home and Student version were used to analyse the data. The data were summarised in frequencies and percentages and then presented in tables, however, seed rate and yield data were subjected to descriptive analysis.

Results and Discussion

Paddy and maize is the major crop being cultivated in the village. Among the paddy, there were more than 50 local cultivars but many of them have been lost with time. Presently, 17 cultivars are being cultivated in Kikruma (Table 1). Maize, garden pea, pyrilla, job's tears, rajma and beans are the common crop cultivated in zabo, but these crops are taken mostly on bunds of the terrace fields. However, under jhum only kharif crops are cultivated, so farmers don't grow garden pea in jhum fields. Respondent listed monkey bean, cabbage, spring onion, garlic, beans, tree tomato, chow-chow, pumpkin, colocasia, brinjal as vegetables cultivated in zabo area, but vegetables like mustard, onion, garlic, tomato and brinjal is not cultivated in jhums. Papaya, banana, guava, plum, peach, pear, lemon, pomelo, pomegranate, gooseberry and wild apple were common fruit planted on bunds or near to ponds in case of zabo but in case of jhum, papaya, banana, guava, plum, peach and pear are grown (Table 2). Further, the respondents revealed that the Common crop, Asian snakehead and snails are reared in ponds dug in the middle of the terraces, whereas cattle, buffalo and goats are commonly reared livestock in the zabo system, whereas under jhum, neither fishes are grown nor livestock is reared.

Table 1: List of paddy (*Oryza sativa*) varieties cultivated in the jhum and terraces in Zabo system

Sl. No	Lowland Paddy Cultivars (Terraces)		Upland Paddy Cultivars (Jhum)
	Sticky	Non-sticky	
1.	Nyode	Tanyekemuga	Thüri (Sticky)
2.	Nyogo	Ribolü	Kumunyothuziri
3.	RhunyoKüzü	White tanie	Richolü
4.	PvakhrüMunyo	Red tanie	Rüli
5.	DzüchoMünyotanie	Chide tanie	Caha (black rice)
6.	Kumunyode	Thüvüri	

Table 2: List of other crops, their local cultivars and enterprises taken in the zabo and jhum system

Sl. No	Plant/livestock/fish	Scientific Name	Local Name of cultivars	Zabo	Jhum
Cereals and pulses					
1.	Maize	*Zea mays*	Kotho, Methohubo,TüphreSako, KhünelüSako, TüphreSako (Non sticky) and Tieciesako and Tsakotsa (Sticky)	√	√
2.	Garden pea	*Pisumsativum*	Motor	√	-
3.	Pyrilla	*Perillafrutescens*	Chitsü, Küna	√	√
4.	Job's tears	*Coixlacryma*	Küse	√	√
5.	Rajma	*Phaseolus vulgaris*	Batüse	√	√
6.	Beans	*Vignaspp*	Kürhise, Kütirese Batüse Tikünalüse	√	√
Vegetables					
7.	Mustard	*Brassica juncea*	Gakri	√	-
8.	Spring onion	*Allium cepa*	Thomüra	√	-
9.	Garlic	*Allium sativum*	Chimuri	√	-
10.	Local garlic	*Allium tuberosum*	Khova	√	√
11.	Brinjal	*Solanum*	Khadeshie	√	-
12.	Chilli	*Capsicum spp.*	Rajathise (Boronithise), Nhathise, Pethimithise, Thisesebu, Thisekütü	√	√
13.	Tomato	*Lycopersiconesculentum*	Ga-o Borase	√	-
14.	Tree tomato	*Solanum betaceum*	SüBongalse	√	√
15.	Cabbage	*Brassica oleraceae*	Kobi, Kobinirhu	√	√
16.	Potato	*Solanum tuberosum*	Galu	√	√
17.	Cucumber	*Cucumissativus*	Zotutuse	√	√
18.	Sponge gourd	*Luffa cylindrica*	Rasuse	√	√
19.	Chow-chow		Skuish	√	√
20.	Pumpkin	*Cucurbita spp.*	Lemu	√	√
21.	Colocasia	*Colocasiaesculenta*	Bitha, Bide, Bicho, Khusobi Thükhrübi, Biyhozho, Bile, Bivene, Tokuswulubi	√	√
22.	Monkey bean	*Parkiaspeciosa*	Yangchak	√	√
Fruits					
23.	Papaya	*Carica papaya*	Thünothipuse	√	√
24.	Banana	*Musa spp.*	Lümüngase	√	√
25.	Guava	*Psidiumguajava*	Maduram	√	√
26.	Plum	*Prunusdomestica*	Agashü,	√	√
27.	Peach	*Prunuspersica*	Chühoshie	√	√
28.	Lemon	*Citrus limon*	SüsuseKükhro	√	-
29.	Pomelo	*Citrus grandis*	Süsusekuzho	√	-
30.	Pomegranate	*Punicagranatum*	Turashie	√	-
31.	Gooseberry	*Phyllanthusemblica*	Nyikhroshie	√	√
32.	Wild Apple	*Docyniaindica*	Tsüphose	√	√
Livestock					
33.	Cattle	*Bosindicus*	Metho	√	-
34.	Buffalo	*Bubalusarneecarabanesis*	Külie	√	-

35.	Goat	*Capra aegagrushircus*	Temvü	√	-
Fishes					
36.	Common carp	*Cyprinuscarpio*	Common carp	√	-
37.	Asian Snakehead	*Channastriatus*	Pfutho	√	-
38.	Snail	*Viviparus(Bellamyia) bengalensis&Pomaceacanaliculata*	Nula	√	-

Paddy being the staple food of the region is a major component of both zabo and jhum farming system. So, data pertaining to cultural practices and yield of paddy have been recorded and presented in the present study. Time of sowing paddy is mid-March to mid-April in both zabo and jhum system. Weeding is only intercultural operation carried in both the system. Transplanting is done during the month of June in zabo system but in jhum, paddy is sown directly. Plant to plant and row to row distance is maintained 10 × 10 and 10 × 15 cm and depth of water column are maintained as 10 to 20 cm in case of zabo, however, seeds are broadcasted in case of jhum and it is entirely rainfed. The fields are kept dried, as no water column is maintained (Table 3). The lower productivity of paddy under jhum is attributable to non-adoption of efficient rainwater management, weeding, improper sowing, and lack of sound plant protection measures (Rathore, 2008).

Table 3: Cultural practice adopted in paddy cultivation under zabo and jhum

Cultural practices	Parameters	Distribution of respondents	
		Zabo (%)	Jhum (%)
Time of sowing	March-April	100.00	100.00
Intercultural operation	Weeding	100.00	100.00
Time of transplanting	June	100.00	Directly sown
Spacing	10x10 cm	37.50	Seed is broadcasted, so no proper spacing is maintained
	10x15 cm	62.50	
Water depth	10 cm	32.50	Rainfed, No water column is maintained
	20 cm	67.50	

In Nagaland, area under Jhum is about 56.50% and contributes 49.26% to total rice production. Studies show that the productivity of paddy in wet Terrace cultivation is more than the Jhum cultivation (Kithan, 2014). Similar observations were also recorded in the present study.The seed rate varied from 15.00 to 60.00 kg with an average of 37.90 (±2.49) kg in zabo and it varied from 25.00 to 75.00 kg with an average of 46.63 (±1.78) kg in jhum. The yield of paddy in zabo terraces varied from 1400 to 2500 kg/ha with an average of 1950.00 (±50.0) kg/ha, whereas it is found to be less than 1.0 ton/ha in Jhum. The fish yield in the small ponds dug in the middle of paddy field under paddy cum fish culture ranged from 40.00 to 80.00 kg with an average of 61.75 (±1.79) kg/ha (Table 4).

Table 4: Seed rate, yield of paddy and fish under zabo system

Variable	Zabo			Jhum		
	Average (Kg/Ha)	Min. (Kg/Ha)	Max. (Kg/Ha)	Average (Kg/Ha)	Min. (Kg/Ha)	Max. (Kg/Ha)
Seed rate of paddy	37.90	15.00	60.00	46.63 (±1.78)	25.00	75.00
Paddy yield	1950.00 (±50.0)	1400.00	2500.00	1570 (±0.05)	950.00	2600.00
Fish yield	61.75 (±1.79)	40.00	80.00	-	-	-

The study showed that majority of farmers of Kikruma area were highly satisfied with zabo system than jhum. Among the respondents, 85% farmers had found zabo as more sustainable practice than jhum. They also found zabo less labour intensive (62.5%) and highly tolerant to climatic vulnerability (75%). Zabo also helps in preserving soil fertility (60%), more remunerative (72.5%), eco-friendly (72.5%), nondestructive (70%) and controls soil erosion (75%). 77.5% of the respondents found that the zabo helps in maintaining ecological balance (Table 5). Zabo found to be more remunerative, as it provides better income opportunity as overall yield is higher and makes available additional income from other subsidiary enterprises like livestock farming and fisheries.

Table 5: Perception of the farmers regarding zabo and jhum farming system

S. No	Parameter	Zabo		Jhum	
		Frequency	Percent	Frequency	Percent
1.	Sustainable	34	85.00	6	15.00
2.	Less labour requirement	25	62.50	15	37.50
3.	Tolerant to climatic vulnerability	30	75.00	10	25.00
4.	Preserves soil fertility	24	60.00	16	40.00
5.	Remunerative	29	72.50	11	27.50
6.	Eco-friendly	29	72.50	11	27.50
7.	Non-destructive	28	70.00	12	30.00
8.	Reduces soil erosion	30	75.00	10	25.00
9.	Creates ecological balance	31	77.50	9	22.50

Conclusion

Zabo is a distinctive and time-tested farming system developed by the farmers of Kikruma range and now being practised in other Chakhesang and Angami regions of Phek and Kohima districts of Nagaland. The practice and knowledge about the sustainable utilization of land, water, soil, forest, livestock and fish resources are unique in zabo. From the study, it can be concluded that zabo farming system is more remunerative, preserves soil fertility, eco-friendly as against to jhum system, which leads to the wider destruction of the area, accelerates soil erosion and creates a severe ecological imbalance.

Acknowledgement

Authors express their thankfulness for the help and support rendered by the people of Kikruma village of the Phek, Nagaland. We also acknowledge the help and support rendered by the Director, ICAR-NRC on Mithun, Nagaland in carrying out the study.

References

Kithan, L.N. 2014. Indigenous form of Paddy cultivation in Terrace and Jhum fields among the Nagas of Nagaland. International Journal of Scientific and Research Publications, 3(4) 2014. www.ijsrp.org/ research-paper-0314/ijsrp-p2751.pdf downloaded on 28th Feb, 2017

Pulamte, L. 2008. Indigenous agricultural systems of North-east India; India, Science and Technology: 2008, published by National Institute of Science, Technology and Development Studies (NISTADS), CSIR, New Delhi. http://www.nistads.res.in/indiasnt2008/t6rural/ t6rurl8.htm downloaded on 15th October 2010.

Rathore, S.S.2008. Paradigm shift for enhancing rice productivity in Nagaland: Existing practices and their refinement. Himalayan Ecology, 16(2): 17-25.

Sharma, U.C., Prasad, R.N. and Sonowal. 1994. An indigenous technique of soil and water conservation in north-eastern region. The Zabo system of farming. Soil and Water Conservation Challenges and Opportunities. In: Proceeding of 8th, ISCO conference (Bhushan, L.S., Abrol, I.P. and Rama Mohan Rao, M.S.Eds). Oxford and IBH, publication Co. Pvt. Ltd., New Delhi (India). pp. 969-975.

Sharma, U. C. and Sharma, V. 2003. The "Zabo" soil and water management and conservation system in north-east India: tribal beliefs in development of water resources and their impact on society – an historical account of a success story. In: Proceedings of the UNESCO/ IAHS/IWHA symposium held in Rome, December 2003. IAHS Publ. 286, 2004, pp 184–192.

Singh, A.K. 2007. Indigenous water management system by the farmers of northeastern hill region, Leisa India, March 2007. http:// www.agriculturesnetwork.org/magazines/ india/1-farmers-coming-together/indigenous-water-management-system-by-the-farmers/at_download/article_pdf downloaded on 23rd September, 2016.

Singh, R.K., Singh, V. and Rajkhowa, C. 2012. Zabo: A Traditional Way of Integrated Farming. In: Resilient Shifting Cultivation: Challenges and Opportunities (Deka, B.C., Patra, M.K., Thirugnanavel, A., Chatterjee, D., Borah,Tasvina R. and Ngachan, S.V.Eds). ICAR Research Complex for NEH Region, Nagaland Centre, Medziphema, Nagaland. pp 114-117.

❑❑❑

Chapter 8

Land-use Planning and Water Harvesting in Dorika Watershed using GIS and Remote Sensing Technique-An Approach for Alternative Land Management System on Watershed Basis

Arunima Gogoi, M.C Talukdar, A. Basumatary, U. Baruah

Department of Soil Science, Assam agricultural University, Jorhat-785013,
E-mail : arunimagogoi123@yahoo.co.in

ABSTRACT

The present study was undertaken to prepare land use plan for Dorika watershed of Sivasagar district in Assam based on soil site suitability evaluation of crops. The watershed lies between $26^{0}48^{\times}$ to $27^{0}03^{\times}$ N latitude and $94^{0}27^{\times}$to $94^{0}52^{\times}$ E longitude. The watershed boundaries were delineated using Survey of India toposheet (1:50,000) no. 83J/13, 83J/9, 83I/12 and I/8 and. R2 L4 MX Remote sensing data. Generally, the soils under this watershed were found to be strongly acidic, high in organic carbon content, low in CEC, medium to high base saturation. Texture varies from coarse to fine texture. Soil pH, drainage and nutrient constraints were the limiting factors for potential utilization of the land resources. Soil site suitability of crops was evaluated for crops, such as rice, mustard, potato, tea, etc. Several crops are recommended depending on their suitability for the post rainy season in the soils which are kept fallow in the present land, use system. Rice is the main crop under this watershed and cultivated during the rainy season for which soils are moderately and marginally suitable. In present study, it was observed that low precipitation in early crop growth stage or growing cycle during *rabi* crops is one of the major constraints. To deal with such situations or dry spells, for cultivation of *rabi* crops, it becomes necessary to harvest rainwater in water harvesting structures for life saving or supplementary irrigation. So, harvesting or storing of rainwater appears to be an alternative proposition. On the basis of remote sensing data and ground truth of river water courses, three nos. of water harvesting structures on Dorika river have been proposed. Considering suitability criteria, site specific cropping plan for different seasons were prepared. The study also covers identification of water harvesting structure at suitable sites; out of three feasible water harvesting structure identified, one water harvesting structure could store approximately 11.61 hectare meter of water, which can provide life saving irrigation during *rabi* seasons for an area of 193 ha land for potato/rapeseed. 232.2 ha for cabbage, 77.4 ha for tomato/direct seeded ahu rice and 50.4 ha for transplanted ahu rice. Results of this study revealed that the study area under rice fallow cropping system could be modified into double cropping system with the use of proper fertility management and life saving irrigation.

Keywords: Land-use planning; soil site suitability; watershed; water harvesting structure.

Introduction

Sivasagar district of Assam occupies an area of 2, 66,800 ha constituting 3.40 per cent of the states geographical area (Dutta, 2009). Dorika is one of the major rivers in Sivasagar district. Dorika watershed covers an area of 27,352.5 ha. The watershed atlas code of Dorika watershed is 3B3D1 (All India Soil and Land-use Survey, 1990). The river Dorika originates from the foothills areas of Assam bordering Nagaland flows though Sivsagar district and merge into the river Dikhou near Brahmaputra. Finally, falls into the river Brahmaputra. More than 90 per cent people of the watershed are dependent on agriculture but have poor production due to poor management of soil that necessitates the study for scientific land-use planning. Therefore, proper land-use planning on the basis of land evaluation is of paramount importance for rational utilization of land resources. This watershed has a niche of different physiographic units and the soils have different inherent characteristics. Most of the cultivated areas under this watershed remain inundated for most part of the rainy season. Rice is cultivated during the rainy season and thereafter, most of the cultivable lands remain fallow. A critical appraisal of land resources and then allocating them to the various cropping system in both rainy and post rainy seasons based on their potential will improve the cropping intensity and compensate the crop loss suffered during rains and floods (Vadivelu *et.al.*, 2003). In such attempts, the soil maps prepared with sufficient ground truth play a vital role in fitting the site specific cultivation. Therefore, the present investigation was taken up to evaluate the soils of Dorika watershed for growing various crops, particularly introducing cultivation of some important crops during rabi season in the dominantly rice alone cropping system and to provide scope for water harvesting in Dorika watershed using GIS and Remote Sensing Technique.

Materials and Methods

The information pertaining to the study area as well as the materials and methods utilized in the present investigation is described under the following heads:

Site description

The Dorika watershed of Sivasagar district is a part of the Upper Brahmaputra Valley Zone of Assam. This watershed has an area of 273.5 sq km comprising topography of hill to alluvial plain. The drainage network of this watershed comprises Namsai Nai, Taola jan, Dorika River, Chataijan, Nimaijan, Nimonagarh , Timuk Nai and Dillih Nai.

The watershed area represents the plains of the Brahmaputra valley. The alluvium deposited at the foothill slopes forms the piedmonts. Geology of the area is alluvium of Pleistocene (Wadia, 1966) and recent periods in the plains and tertiary sedimentary rocks in the hill. The soil moisture of this study area is Udic. and *hyperthermic* regime.

The northern part of this watershed particularly along the river Brahmaputra is often subjected to flooding during the rainy season. A major part of the watershed is often remains inundated for the most part of the rainy season. The drainage condition is poor.

Climate

The climate of this area is humid subtropical with an average annual rainfall of 2,334 mm. The soil moisture of this study area is Udic and the soil temperature regime is hyperthermic.

Collection of soil samples & Laboratory analysis

Nine profile (depth wise) samples representing different physiographic units were selected and horizon-wise soil samples were collected from each of the selected pedon. Morphological features of each pedon were described with standard procedure (All India Soil and Land-use Survey Organization, 1971).

Morphological features of each pedon were studied in the field following standard procedure (All India Soil and Land-use Survey Organization, 1971). The standard methods were adopted for physico-chemical analysis of the soil. Soil classification was carried out on the basis of morphological and physico-chemical properties of soils, following *'Keys to Soil Taxonomy'* (Soil Survey Staff, 2006).

Delineation of Dorika watershed

The boundary of Dorika watershed was delineated based on Survey of India toposheet No. No. 83J/13, J/9 and 83I/12 on 1:50,000 scale and IRS, Resourcesat- 2 L4 FX remote sensing data (1:50,000 scale).

Results and Discussion

Digital data of IRS, R-2 L4FX data of November and December, 2011 with a spatial resolution of 5.8 m was used and geocoded using TNT mip (Map image processing) software with reference to toposheet. Geocoded FCC was visually interpreted and in conjunction with survey of India toposheet No. 83J/13, J/9 and 83I/12 on (1:50,000 scale). Remote sensing data (Resources at- 2, L4 FX) were geo referenced and digitized to delineate the watershed boundary based on drainage network, water flow path and ground truth verification. Digital Elevation Model (DEM) map was also used to see the water flow path. Finally watershed boundary area was extracted from the imagery with the help of GIS tools. (Fig. 1 & 2).

Preparation of physiographic map

Physiographic map of Dorika watershed was prepared based on Survey of India toposheet No. 83J/13, J/9 and 83I/12 (1:50,000 scale) in conjunction with IRS Resources at- 2, L4 FX data. The Remote sensing data were geo-referenced and digitized to prepare the physiographic map. Visual and digital interpretation was used to delineate the watershed boundary and physiographic units within the boundary. Results were validated based on the ground. Four distinct physiographic units were identified within the watershed boundary, which includes alluvial plain, recent flood plain, piedmont plain and hill (Fig. 2). Alluvial plain appeared light green in colour with coarse texture and non-linear pattern. It was found that 193 sq km area was occupied by alluvial plain and 44 sq km area occupied by recent flood plain. About 31 sq km area was occupied by piedmont plain. Fine texture with dark uniform red to medium red tone indicated hills in the image. Another 4.97 sq km area was found to be under hills.

Soils

Four distinct physiographic units were recognized in the studied area which included alluvial plain, recent flood plain, piedmont plain and hill soils. Nine soil site locations were identified under this watershed, representing each physiographic unit. Alluvial plain soil (Gormur, P2; Rajabari, P3; Moran gaon, P4; Luthurichetia gaon, P5 and Banmukh, P7). The textures of the soils were sandy loam, silt loam, clay loam, silty clay to clay. The soils of recent flood plain soil (Dorikaghat, P6; Chutia gaon, P8; Mogona gaon, Panbecha P9) which situated in northern side of river Dorika, are silty loam in texture. Whereas, the soils of Santak are the south-eastern part of this watershed lies near the Naga Hills and texture varies from sandy loam to loamy sand. Soils under this watershed were found to be strongly acidic, high organic carbon content, low in CEC, medium to high base saturation and texture varies from coarse to fine texture. The pedons (P1, P2, P3, P4, P5, P6, P7 and P9) of Dorika watershed were classified as Inceptisols .The temperature regime of this soil was hyperthermic and pedon P1, P2, P3, P4, P6, P7 and P9 qualified for udic soil moisture regime. So, these soils were classified as Udepts as suborder level. These soils qualified Dystrudepts at great group level. The soils of P8 (Chutia gaon) belonged to Entisols, as these soils did not have any diagnostic horizon other than ochric. These soils were saturated with water during some parts of the year and had low chroma (1-2) accompanied by mottles. These soils were classified as Aquents suborder and Endoaquents at great group level. Further, the soil represented the central concept of subgroup. So, these soils were placed in *Typic Endoaquents* at subgroup level.

Soil site suitability evaluation of crops

The suitability of the soils for growing rice, wheat, maize, mustard, potato, banana and tea was evaluated by Sys method (Sys *et. al.*, 1993).The soil characteristics were matched with the requirements of the crops at different limitation levels. The suitability classes and subclasses were decided by most limiting soil characteristics. After the evaluation, the soil units comprising two or more soil series were allotted to these crops for which they are highly suitable (S1) and moderately suitable (S2). The crops, which were in association with marginally suitable (S3) and presently not suitable (N1) classes, are not recommended because their cultivation will not be economical. Thus, the site specific cropping plan was prepared for identified location of Dorika watershed, based on soil characteristics and limitation of soil site suitability.

Soil site suitability evaluation for major crops

Cereals

***Winter* rice (Sali):** As per Sys's (Sys *et al.*, 1993) soil site suitability assessment revealed that soils of piedmont plain were not suitable (N2), alluvial plain soils were marginally suitable (S3) and recent flood plain soils (S2) were moderately suitable for *winter* rice cultivation (Table 1). The soils of Santak (piedmont plain), Gormur and Rajabari (alluvial plain) were found to be unsuitable for *winter* rice cultivation due to limitation of flooding and drainage. Among Sys parameters, flooding and drainage were found to be the major constraints controlling the suitability class. But, in *kharif* season, rice is the major crop cultivated in all soils of the study area. In Assam, more than 80 per cent of annual rainfall used to receive during this period, which coincides with *winter* rice and that reflects the average yield efficiency percentage (AYEP) of *winter* rice. The landscape of piedmont plain and alluvial plain soils was not experienced under flooding and all the rice cultivated soils used to remain saturated due to continuous rainfall. Hence, flooding and drainage could not be considered as major constraints. In addition, *winter* rice showed the average yield efficiency percentage (AYEP) of 169.81 for piedmont plain, 205.65 for alluvial plain and 184.76 per cent for recent flood plain. *Winter* rice was found to be climatically highly suitable for the study area. So, the soils of Dorika watershed could be modified into moderately suitable (S2) to marginally suitable (S3) in alluvial plain and marginally suitable (S3) in piedmont plain soil. In the study area, it was also observed that the alluvial plain soils were more suitable for rice crop than the recent flood plain, followed by the piedmont plain soil. Yield losses were observed in recent flood plain area due to flooding, which contribute lower yield in recent flood plain as compared to alluvial plain.

***Autumn* rice (Ahu rice):** Suitability evaluation for autumn rice (Ahu rice) by Sys *et al.* (1993) method revealed that the all soils under this watershed were marginally suitable for (S3) ahu rice cultivation (Table 1) due to limitation of precipitation in early period of crop growth. Besides this, textural limitation, low pH and unsuitable drainage condition was found as limiting factor for ahu rice cultivation. As per Sys *et al.* (1993), low organic carbon content was found as one of the major constraints. Moreover, AYEP of autumn rice was almost twice (210.79) than that of state average indicating better adaptability of autumn rice in that area. Considering these factors, Sys's suitability class for autumn rice could be modified to moderately suitable in alluvial plain and recent flood plain soil. Piedmont plain soil remained as same as mentioned by Sys's suitability evaluation because of textural limitation.

Wheat: Sys's suitability evaluation for wheat crop revealed that the whole area was unsuitable (N2) for growing of the crop due to low precipitation during early crop growth (Table 1). However, unprecedented rainfall during the month of March and April coinciding for wheat maturity was a serious limitation in this study area. Selection of short duration variety might encounter such situation. Hence, the suitability classes were slightly modified to presently not suitable (N1) for all site of location under the study area. Low precipitation during growing cycle, soil acidity, and texture were found to be key limitations for growing wheat crop in these areas. Similar findings were also reported by Borah (2006) in Kamalabari area of Assam and Baruah (2003), Vadivelu *et al.* (2004a) and Dutta (2009) for some soils of the Brahmaputra valley.

Rabi Maize: As per Sys's soil site suitability, assessment revealed that *rabi* Maize was unsuitable (N2) for cultivation due to low precipitation during growing cycle.

Oilseeds

Rapeseed mustard: The study area was not suitable (N2) for rapeseed/mustard due to low precipitation, high relative humidity, poor drainage and low pH content (Table 1). But considering average yield efficiency percentage (AYEP) of rapeseed mustard, all these Sys's suitability classes were modified to moderately suitable (S2) for all three physiographic unit except the soils of Rajabari (alluvial plain) and Panbecha (recent flood plain).

Vegetables

Potato: Study area was found to be unsuitable (N2) for potato (Table 1). According to Sys limitation, low precipitation is the major constraints for cultivation of potato. But, in the study area, high average yield efficiency percentage (AYEP) was found for potato. It might be due to low precipitation, could not affect the crop severely as high relative humidity and dew could sustain the water for some periods only. So, considering average yield efficiency percentage (AYEP) of potato, Sys suitability class for potato of piedmont plain was modified to highly suitable (S1).

Tomato: Suitability evaluation for tomato as per Sys *et al.* (1993) revealed that the study area were unsuitable (N2) due to low precipitation during growing cycle (Table1). Low precipitation in early crop growth stage, pH and organic carbon is the major constraints that can limit the cultivation of Tomato. By application of lime and organic manure, the suitability class of alluvial plain and recent flood plain could be modified to moderately suitable (S2). Due to textural limitation, the piedmont plain soil is marginally suitable (S3).

Cabbage: Cabbage was not suitable (N2) in all physiographic condition due to the constraints of low precipitation in growing cycle and low pH of the soils of watershed (Table 1) By application of life saving irrigation, and lime the crop suitability could be modified into highly suitable (S1) in piedmont plain soil and moderately suitable (S2) in alluvial plain (Gormur, Moran gaon, Luthurichetia gaon and Banmukh) and marginally suitable in recent flood plain (Dorikaghat, Chutiagaon and Panbecha). In Rajabari, soil of alluvial plain was marginally suitable (S3) due to presence of clay texture.

Pea: As per Sys's suitability assessment, the area was not suitable (N2) for growing pea due to low precipitation during crop growing cycle in this study area (Table 1). But in study area, low precipitation could not hamper the cultivation of those crops rather it was beneficial. pH was found to be one of the major constraints for cultivation of pea. Well-drained soils with high organic matter content allow better root aeration and thereby less disease attack like root rot. High acidity may require liming with provision of life saving irrigation, suitability of pea could be modified to highly suitable in the soils of Rajabari (alluvial plain) and moderately suitable to marginally suitable (S3) for other soils of alluvial plain, recent flood plain and piedmont plain soil with consideration of texture, NPK status and organic carbon.

Plantation crop

Tea: As per Sys's suitability the assessment for cultivation of tea, the piedmont plain soil (P1) was found to be marginally suitable (S3) and limitation approach showed that the alluvial plain soils were moderately suitable (S2) to marginally suitable (S3). The climate was found to be highly suitable (S1) (Table 1). Due to flooding in recent flood plain area with soil limitations of organic carbon, flooding, and poor drainage condition the soils of recent flood plain area was not suitable (N2) for tea cultivation. It was observed that the soils of Santak (piedmont plain) are presently under tea cultivation. It was reported that light texture soil like sandy loam or loamy soil was the best texture for tea cultivation. So, considering textural suitability with application of organic manure, the soils of piedmont plain soil (Santak) could be modified to highly suitable. The alluvial soils (Gormur and Rajabari) could be categorised as moderately suitable and remaining others (Moran gaon, Luthurichetia gaon and Banmukh) were marginally suitable (S3) for tea cultivation. Recent flood plain area was not suitable (N1) for tea cultivation due to flooding. Soil fertility and poor drainage were most limiting factors. By improving fertility status and drainage, most of the studied soils of alluvial plain could be upgraded to moderately suitable (S2) to highly suitable (S1) class.

Banana: Soil site suitability for horticultural crops was also evaluated as per Sys' method. Banana is moderately suitable (S2) in alluvial plain soil of Gormur and Rajabari.

Sugarcane: According to Sys suitability evaluation, the climatic suitability was poor for sugarcane, hence it was found to be unsuitable (N2) for all three physiographic units for sugarcane cultivation (Table 1). But considering other parameters of soil, the suitability of sugarcane in piedmont plain soil could be modified to moderately suitable (S2). Considering, textural suitability, alluvial plain soil was categorized into marginally suitable (S3). Recently, flood plain area was not suitable (N1) due to occurrence of flooding. By application of management practices like irrigation, improving drainage condition and fertility status, the suitability of sugarcane could be improved.

Major constraints

The major constraints of the watershed lie in the physical properties (texture), poor drainage, low precipitation during early crop growth stage or growing cycle of *rabi* crops, acidity and poor fertility. The acidity and fertility limitation could be improved by using appropriate amendments like lime, organic manure and green manure, etc. Most soils of recent flood plain and piedmont plain, owing to light textured could be exploited for atleast two crops during *rabi* and post *rabi* seasons. Low precipitation during early crop growth stages or growing cycle of *rabi* crops could be addressed by increasing soil moisture status either by following moisture conservation technique or through supplementary irrigation.

Proposed Water Harvesting Structure

In present study, it was observed that low precipitation in early crop growth stage or growing cycle during *rabi* crops is one of the major constraints. To deal with such situations or dry spells, for cultivation of *rabi* crops, it becomes necessary to harvest rainwater in water harvesting structures for life saving or supplementary irrigation. So, harvesting or storing of rainwater appears to be an alternative proposition. On the basis of remote sensing data and ground truth of river water courses, three nos. of water harvesting structure (Fig. 3, Fig 4) in the river have been proposed. While doing so, minimum disturbances in built up area, cultivated land and economics were considered. In the present investigation, detail on one proposed water harvesting structure (W1) was discussed (Fig 4.a). The proposed water harvesting structure (Fig 4a, Fig 4b) covers a latitude 27°1'?37" N to 27°2'?9"N and longitude 94°36'?27" E to 94°36'?53"E with an area of 75.34 m^2 (approximate) with a length of 2500 m (approximate). Hence, it will be able to store 19,35,00,000 litres of water. Assuming, total stored 19,35,00,000 litres water or 19.35 hectare meter, and 40% water loss (including evaporation, seepage, infiltration loss, etc.). The remaining of 11.61 hectare meter water can be stored which could be used for life saving irrigation during critical crop growth periods particularly *rabi* crops (potato, cabbage, rapeseed/mustard, pea, tomato, maize, sugarcane, wheat and ahu rice) for irrigation of an area of 193 ha for potato/ rapeseed, 232.2 ha for cabbage, 77.4 ha for tomato/direct seeded ahu rice and 50.4 ha for transplanted ahu rice. Irrigation requirement for *rabi* crops and area covered are given in the Table 3.

Land-Use Plan

Based on soil site suitability of crops, socio-economic conditions, length of growing crop period, occurrence of flood, net return and major constraints of the study area few land-use plan for different physiographic units were suggested. It is expected that implementation of land-use plan would enhance the existing productivity of the study areas. On the basis of soil site suitability criteria and limitation (Table 1) site, specific cropping plan was prepared for different seasons (Table 2). Various cropping system like ahu rice-sali rice-rapeseed/ mustard, ahu rice-sali rice-rabi vegetables can be adopted in study area. The areas not found suitable for growing any crops can be utilized for other livelihood options.

Conclusion

The area experiences sufficient rainfall with length of growing period of 290 days and optimum temperatures suitable for a number of crops. But, pH, drainage, physical texture, nutrient constraints become the limiting factors for potential utilization of the land resources. Some agricultural crops like potato, mustard, maize, pea, green gram, french bean can be grown in these soils with liming and fertilizer management. Besides this, the water stored in the soils is in appreciable amount which can be utilized for intensification during Rabi under rainfed condition. The water stored in the water harvesting structure can be utilized for life saving irrigation during critical periods, particularly for growing wheat and potato. Results of this study revealed that the study area under rice fallow cropping system could be modified into double cropping system with the use of proper fertility management and life saving irrigation.

Acknowledements

The authors are thankful to the Assam Agricultural University, Jorhat for providing research facilities to carry out the experiment.

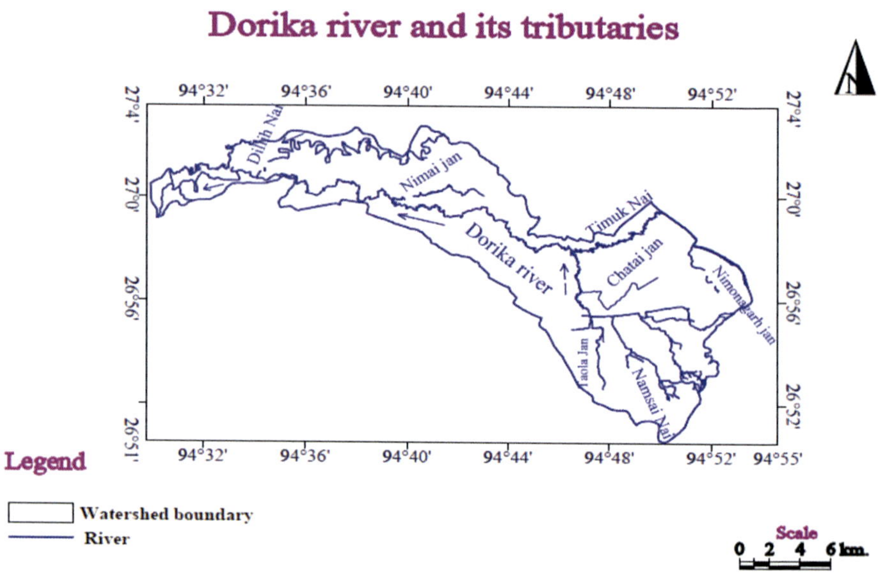

Fig. 1: Delineation of Dorika watershed

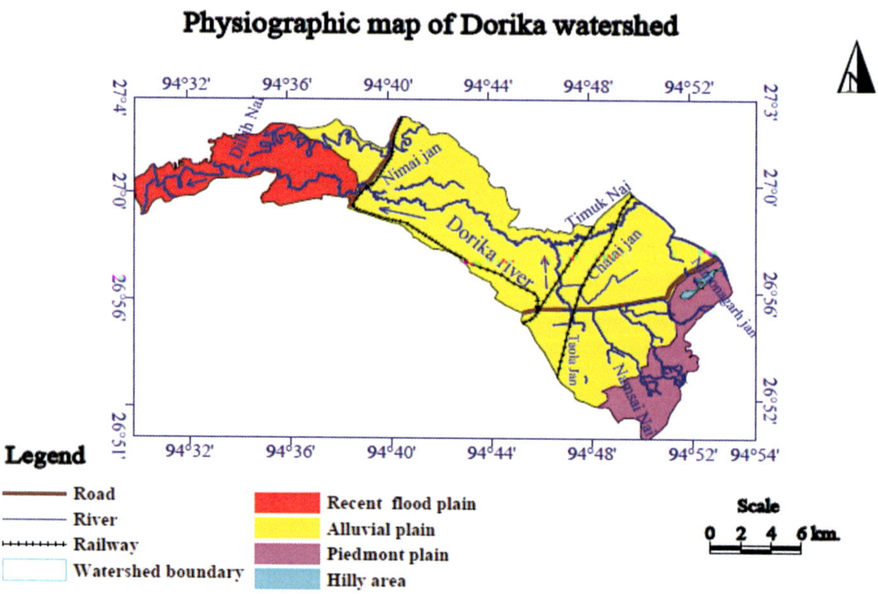

Fig. 2: Physiographic map of Dorika watershed

Table 1: Suitability evaluation of Dorika watershed

Location No.	Sali rice	Ahu rice	Wheat	Maize	Potato	Cabbage	Pea	Tomato	Rapeseed/ mustard	Tea	Sugarcane	Banana
Piedmont plain												
P1(Santak)	S3*wsf*	S3 *cwsf*	N1*csf*	S3 *csf*	S1 *c*	S2 *cf*	S1*cfs*	S3 *cs*	S2 *cf*	S1*sf*	S2 *cfs*	S3*fsc*
Alluvial plain												
P2(Gormur)	S3*wfs*	S2 *cwsf*	N1 *csf*	S3*cf*	S1 *cf*	S2*cf*	S2*cf*	S2*cf*	S2*cfs*	S2*cwf*	S3*cws*	S2*fwc*
P3(Rajabari)	S3*wfs*	S2 *cwsf*	N1*csf*	S3 *csf*	S3 *cfs*	S3*csf*	S2 *cf*	S2 *cfs*	S3 *cfs*	S2 *cwfs*	S3 *cws*	S2 *fwc*
P4(Moran gaon)	S2*wfs*	S2 *cwsf*	N1*csf*	N1 *cf*	S2 *cf*	S2 *cf*	S3 *cf*	S2 *cf*	S2 *cfs*	S3*cwf*	S3 *cws*	S3 *fwc*
P5(Luthurichetia gaon)	S2*wfs*	S2 *cwsf*	N1*csf*	S3 *cf*	S2 *cf*	S2 *cf*	S3 *cf*	S2 *cf*	S2 *cfs*	S3 *cwf*	S3 *cws*	S3 *fwc*
P7 (Banmukh)	S2*wfs*	S2 *cwsf*	N1*csf*	S3 *cf*	S1 *cf*	S3 *cf*	S2 *cf*	S2 *cf*	S2 *cfs*	S3 *cwf*	S3 *cws*	S3 *fwc*
Recent flood plain												
P6(Dorikaghat)	S2*wfs*	S2 *cwsf*	N1*cwsf*	S3 *cwf*	S1 *cwf*	S3 *cf*	S2*cf*	S2 *cf*	S2*cf*	N1*fw*	N1 *cws*	N1 *fwc*
P8(Chutia gaon)	S2*wfs*	S2 *cwsf*	N1*cwsf*	S3 *cwf*	S2 *cwf*	S3 *cf*	S3*cf*	S2 *cf*	S2*cf*	N1 *fw*	N1 *cws*	N1 *fwc*
P9(Panbecha)	S2*wfs*	S2*cwsf*	N1*cwsf*	S3 *cwf*	S3 *cwfs*	S3 *cf*	S3*cf*	S2 *cf*	S3 *cfs*	N1*wfs*	N1 *cws*	N1 *fwc*

Note: c: limitation of climate; w: limitation of wetness; f: limitation of soil fertility; s: limitation of soil physical properties

Table 2: Preparation of site specific action plan

Location No.	Autumn (February-June)	Winter (July-November)	Summer (November- March)	Year round plantation/ horticultural crops
Piedmont plain				
P1(Santak)	-	Sali rice	Potato/Cabbage/ Maize	Tea/Sugarcane
Alluvial plain				
P2(Gormur)	Ahu rice	Sali rice	Potato/Rapeseed /Mustard/Pea/Cabbage/Tomato	Tea/Banana
P3(Rajabari)	Ahu rice	Sali rice	Pea/Tomato/Maize	Tea/Banana
P4(Moran gaon)	Ahu rice	Sali rice	Potato/Rapeseed /Mustard /Cabbage/Tomato	-
P5(Luthurichetia gaon)	Ahu rice	Sali rice	Potato/Rapeseed /Mustard/ Cabbage/Tomato	-
P7(Banmukh)	Ahu rice	Sali rice	Potato/Rapeseed /Mustard/Pea/ Tomato/Maize	-
Recent flood plain				
P6 (Dorikaghat)	Ahu rice	Sali rice	Potato/Rapeseed /Mustard/Pea/ Tomato	-
P8(Chutia gaon)	Ahu rice	Sali rice	Potato/Rapeseed /Mustard/Pea/ Tomato	-
P9(Mogona gaon, Panbecha)	Ahu rice	Sali rice	Tomato	-

Table 3: Irrigation water requirement of crop

Crop	Amount of water required for irrigation (mm)	Irrigated water required (m)	Per ha irrigated water required (liter)	One irrigation (liter)	Area covered (ha)
Potato	180	0.18	1800,000	6,00,000	193.5
Rapeseed	60	0.06	600,000	6,00,000	193.5
Cabbage	100	0.10	1000,000	5,00,000	232.2
Tomato	300	0.30	3000,000	15,00,000	77.4
Direct seeded Ahu rice	308	0.308	3080,000	15,40,000 (50% of irrigation water)	77.4
Transplanted Ahu rice	475	0.475	4750,000	2375000(50% of irrigation water)	50.4

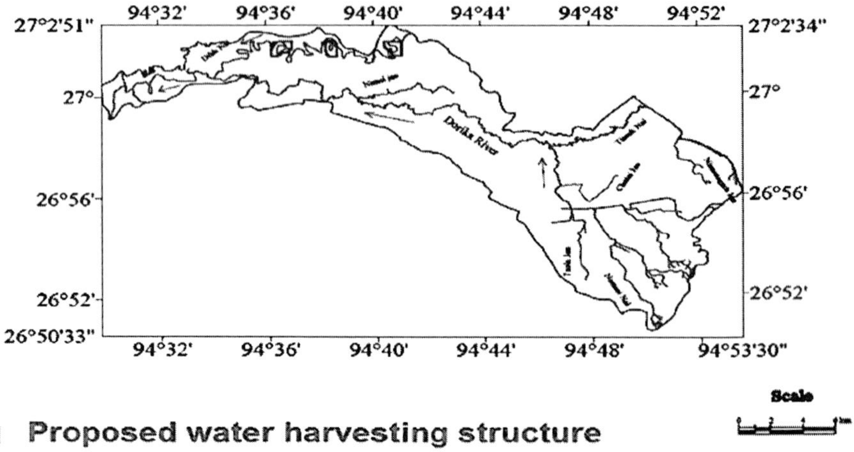

Fig. 3: Proposed three no. of water harvesting structure

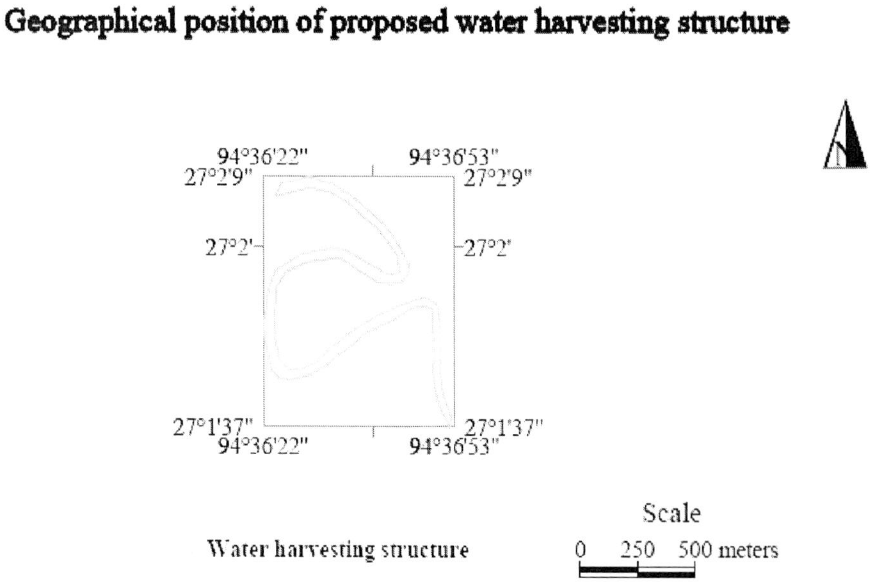

Fig. 4. a. Geographical position of proposed water harvesting structure

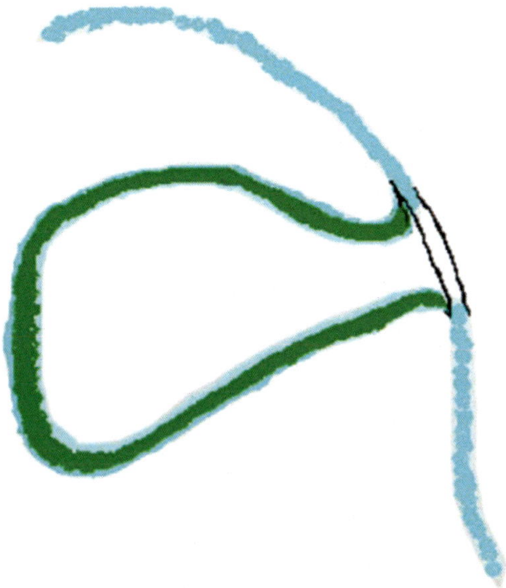

Fig. 4. b. Geographical position of proposed water harvesting structure with river bunding

References

All India Soil and Land-use Survey.1990. *Watershed Atlas of India* (on 1:1 million scale).

All India Soil and Land-use Survey Organization.1971. *Soil Survey Manual,* IARI, New Delhi.

Anonymous.2012. Profile of Agri-Horti sector of Assam. Land Utilization Statistics, Directorate of Economics and Statistics in Assam. www.agriassam.in/agri Horti_profile/Profile_of Agri Horti sector_of Assam_June 2012' pdf.

Deka, Bipul; Baruah, T.C., Dutta, Marami and Karmakar, R.M. 2009. Landscape Soil Relationship and Pedogenic Evaluation of Soils in Ghiladhari Watershed of the Brahmaputra Valley of Assam. *J. Indian Soc. Soil Sci.* 57(3): 245-252

Dutta, M. 2009. Land use planning of Tipukjan watershed of Sivasagar district using GIS and remote sensing technique. Ph.D. Thesis, AAU, Jorhat

Jenny, H. 1931. Behaviour of potassium and sodium during the process of soil formation. *Univesity of Missouri Agricultural Experimental Station Research Bulletin,* pp. 162.

McCloy, K.R. 1995. Resource management information system, processes and practice, Taylor and Francis, U.K.

Sarkar, D., Baruah, U., Gangopadhyay, S.K.; Sahoo, A.K. and Velayutham, M. 2002. Charactrization and classification of soils of Loktak catchment area of Manipur for sustainable land-use planning. *J. Indian Soc. Soil Sci.* **50**: 196-204.

Sen, T.K.; Dubey, P.N.; Nayak, D.C.; Baruah, U.; Bhattacharyya, T.; Maji, A.K. and Velayuthum.2003. Soil resource information for agricultural planning and development of Assam. *Agropedology* **13**: 50-59.

Soil Survey Staff. 2006. Keys to Soil Taxonomy. United States Department of Agriculture, Natural Resources Conservation Service. Washington, D.C.

Sys, C., Van Ranst, E. and Debaveye, J. 1993. Land Evaluation, Part II, Methods in Land Evaluation. Agricultural Publication No.7, General Administration for Development Co-operation, Belgium.

Vadivelu, S., Baruah, U., Bhaskar, B.P., Thampi, J., Sarkar, D., Walia, C.S., Nayak, D.C. and Maji A.K. 2003. A land-use plan for Jorhat district of Assam state. *Agropedology*, **13**:1-10

Wadia, D.N. 1966. *Geology of India.* The English Language Book Society and MacMillan Co. Ltd., London, pp. 76-92.

❑❑❑

Chapter 9

Effect of Diverse Cropping Practices on Runoff and Soil Loss in the Valley-and Foot-Hill Region of Arunachal Pradesh

Saroj Kumar Pattanaik[1*], Ajai Kumar Pandey[1], Prankanu Debnath[1], Balaram Panigrahi[2], and Dileep K. Panda[3]

[1]College of Horticulture and Forestry, Central Agricultural University, Pasighat, Arunachal Pradseh-791102, [2]College of Agricultural Engineering and Technology, O.U.A.T., Bhubaneshwar, Odisha, [3]Indian Institute of Water Management, Chandrasekharpur, Bhubaneshwar, Odisha, *Email: saroj_swce@rediffmail.com

ABSTRACT

A field experiment was conducted during 2013-2014 in the farm of the College of Horticulture and Forestry, Central Agricultural University, Pasighat, Arunachal Pradesh for evaluating the different cropping practices in retarding runoff and soil loss under land slope varying from 2-3%. The average annual rainfall of the study area is about 4510 mm with erratic annual distribution. Effects of five types of cropping practice-slope treatments on conservation of runoff and soil loss were studied. The treatment T_4 (maize at 3% slope) resulted highest amount of runoff i.e. 2198.4 litres, whereas, the treatment T_6 (cowpea at 3% slope) resulted lowest runoff of 1120.5 litres. The same trend was observed for soil loss also. The highest sediment yield was found for the treatment T_4 *i.e.* 865.8 grams and the lowest for the treatment T_6 *i.e.* 289.5grams, which are significantly different from each other. The trend of the results of soil sample nutrient analysis showed that higher content of N, P, and K in lower part of the runoff plots in comparison to their middle and uppers parts. This is due to the removal of nutrient from upper part to lower part of the land along with runoff water. It is found that maize-cowpea inter-cropping at 2:1 ratio resulted in 38.6% less runoff and 30.7% less sediment loss. The farmers usually cultivate maize as one of the remunerative kharif crop. This crop alone is not able to reduce soil erosion under this agro-climatic condition. Inter-cropping of traditional maize with cowpea can reduce the soil erosion considerably in the slopping lands. Hence, the farmers can be suggested not to go for single crop like maize instead to go for inter-cropping of maize and cowpea.

Keywords: Sloping land, runoff loss, sediment yield, inter-cropping.

Introduction

Runoff formation and subsequent sediment and nutrient losses are caused by many anthropogenic factors, such as faulty cultivation practices, faulty cropping pattern and adoption of shifting cultivation practices in slopping lands. In Arunachal Pradesh, the cultivated lands are located in hill slopes, valleys and foot hill region. The slopes of valleys and foot hills regions vary from 2% to 10%. Generally, two types of cropping patterns are prevalent *viz.* mono-cropping and mixed cropping. The former is practised in settled areas, whereas the latter is practised in settled as well as jhum (Shifting cultivation) areas. *Kharif* is the main cropping season and maize is one of the major crops sown by the farmers. However, maize is considered as soil erosion permitting crop. Growing of two or more crops in mixture with maize (Zea mays L) provides better canopy over and obstructs overland flow. Dense foliage of soil erosion resistant leguminous crops like cowpea (*Vigna unguiculata*) and dry bean (*Phaseolus vulgaris*) prevent soil particle detachment as the raindrops cannot hit soil surface directly.

Over the past decades, numerous studies have been conducted to investigate runoff generation mechanisms worldwide (Nett and Hendley, 2002). Studies are also carried out on mixed cropping of maize with rice (*Oryzasativa*) which revealed that runoff and soil loss was reduced significantly compared with maize along the slope. Rice and maize based inter-cropping systems, such as rice+groundnut/ soyabean in 3:1 row ratio and maize +groundnut/soyabean in 1:1 row ratio performed well and resulted in higher net-returns in upland mid altitude of Meghalaya (Satapathy, 2003).

Plot-scale experimental studies are generally part of broader research projects aimed at improving the understanding of inter-relations between processes involving hydrological, climatic and biological factors (Singh and Singh, 1978). Plot-scale studies have the advantage of allowing for detailed process monitoring at small scale, providing a basic description of the most relevant aspects (Michaelides, *et al.*, 2009). Plot-scale studies are also useful in providing experimental data involving rainfall, surface runoff and soil erosion. These data are used as reference in model formulation, calibration and validation.

Small runoff plots ranging from 1/200th of an acre to 1/50th acre have been widely used during the initial stages of research and are still in use to evaluate the comparative efficiency of individual factors, (such as degree and length of slope, vegetative cover, cropping practices, soil, etc.) recognised to be affecting runoff and sediment production (Michaelides *et al.*, 2009). Since plots are artificially made and small enough for normal field operations, the data collected cannot be safely applied to field conditions and therefore, are limited to only comparative study producing qualitative information. The runoff plots have advantage of maximum similarity with respect to size, shape, slope, soil, *etc.* and consequently, minimum variation of factors which are generally associated with natural watersheds. The control conditions of the plots help in evaluation of the effect of any practice in relatively shorter time. Plot-scale studies are also useful in providing experimental data involving rainfall, surface runoff and soil erosion. These data are used as reference in modelling conception, calibration and validation. However, there can be considerable variability in soil erosion processes, as well as limitations of models attempting to simulate these complexities[7].

In this context, a field study was conducted to evaluate the effect of different cropping practices on the runoff and sediment yield in the agro-climatic conditions of the valley and foothill region of Arunachal Pradesh.

Materials and Methods

Study area

The study area is located at the research farm of the College of Horticulture and Forestry, Central Agricultural University of Pasighat, East Siang district of Arunachal Pradesh. The geographical location is between 27.3^0 to 29.42^0 North latitude and 94.42^0 to 95.35^0 East longitude with an altitude of 155m above MSL. The climate of Pasighat is tropical humid during summer and dry mild winter. The long-term average annual rainfall of the East Siang district is 4510 mm distributed over the year. The rain starts here during the last week of March and continues through September. The period from October to March is a dry period.

The soil is highly porous, having low water holding capacity, high infiltration rate and high bulk density which have adverse effect on soil-water-plant continuum. High permeability of soil causes low nutrient use because of downward movement of available plant nutrients away from the root zone of field crops, resulting in low productivity of crops. The soil is also highly acidic. The organic carbon content of the soil ranges from 3.8g kg^{-1} to 18.3 g/kg^{-1}. Anionic plant nutrients like Nitrogen (N), Sulphur (S), Boron (B) are more prone to leaching losses. Soil in the high rainfall areas are again at risk of downward movement of cationic bases as well as loss of nutrients aided by runoff at even the mild gradient also. The soil of the valley and foot-hill region of Arunachal Pradesh below 20 cm depth from ground is embedded with numerous stones, indicates the importance of the top 20cm fertile soil responsible for the growth of crops. The cultivated lands exist on atleast 2% slope and more. The crops like paddy, maize and vegetables are grown in the region. The physico-chemical properties of the soil of study area are given in Table 1.

Table 1: Physico-chemical properties of the study area.

S.No.	Soil properties	Status
1.	Texture	Sandy loam
	(i) Sand (%)	54.00
	(ii) Silt (%)	27.00
	(iii) Clay (%)	19.00
2.	pH	5.6
3.	Available N (kg/ha)	261.5
4.	Available P_2O_5 (kg/ha)	57.0
5.	Available K (kg/ha)	102.5
6.	Organic carbon (gm/kg)	14.2

The site is selected in the slope range from 2% to 3% representing the most commonly practiced slope ranges by the farmers in the valley and foot-hill region of East Siang district of Arunachal Pradesh. The area was divided into six plots accommodating six treatments as shown in the Figures 3-8 and in Table 2. The runoff plots were prepared after cleaning the site. The land was ploughed and the area was divided into six runoff plots. The size of each plot was 20 m × 5 m as shown in Fig. 1 (FAO). The plots were separated from each with the help of bunds of height one foot. Each plot at its upstream and downstream ends is also bounded by bunds having height one foot. The bunds are covered with 250 GSM polyethylene film. The bunds prevent the outside runoff to enter into each plot. The soil analysis of each plot was carried out by following established procedure (Jackson, 1973). Five soil samples were collected from each part of the plot i.e. upper, middle and lower after the crops were harvested. A composite sample was prepared from these five samples and was analysed for its nutrient status.

Table 2: Treatment details with specification of the biological conservation measures

Treatments	Specification
T_1	Cowpea at 2.5% slope
T_2	Maize + Cowpea at 2% slope
T_3	Maize at 2% slope
T_4	Maize at 3% slope
T_5	Maize + Cowpea at 2.5% slope
T_6	Cowpea at 3% slope

Figure 1. Runoff plots in the study area

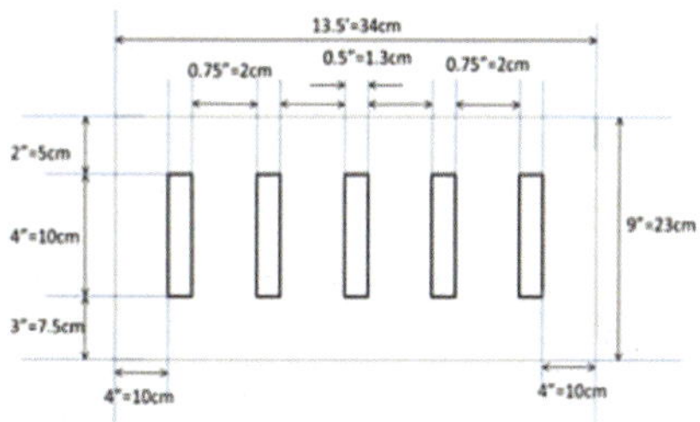

Figure 2. Multislot divisor plate

Figure 3. T_1 : Cowpea at 2.5% slope

Figure 4. T_2: Maize + Cowpea at 2% slope

Figure 5. T_3: Maize at 2% slope

Figure 6. T_4: Maize at 3% slope

Figure 7. T_5: Maize + Cowpea at 2.5% slope

Figure 8. T_6: Cowpea at 3% slope

Cropping Practices

Maize variety RCM-76 a high yielding variety developed at ICAR-RC-NEH, Barapani for North-Eastern region performed well and is normally cultivated by the farmers. Normal sowing time of this crop is March-April and harvested in the month of August-September. Recommended seed rate and spacing for maize is 12-15 kg/ha and 50×20 cm, respectively. After harvesting of cobs, remaining plants are used as green fodder. Taking this into account, maize is taken as a crop in the trial. It is known that maize is an erosion permitting crop. It is inter-cropped with cowpea, which is considered as erosion resisting crop. Hence, the cowpea (*Vignaunguiculata*)variety Khasikanchan is used as a cover crop. Both the crops i.e. maize and cowpea were was sown in the last week of March at a spacing of 45×30 cm and 30×15 cm with 1-2 seeds per hole, respectively. The inter-cropping was carried out at 2:4 *i.e* two rows maize inter-cropped with four rows of cowpea.

Runoff Measurement and Soil Loss Estimation

The multislot divisor unit developed by R.V.Geib in U.S.A. was used for the measurement of runoff from the experimental plots[12]. At the down end of the experimental plots, the multi-slot divisor unit consisting of five numbers of slots of equal dimensions was installed to monitor runoff (Fig. 2). Each slot of the divisor plate acts as a weir, which is frequently used as a device to measure the flow of water i.e. the discharge[6]. The runoff monitored once in 24 hours at 8.30a.m. (IST) by measuring the depth of runoff collected in each collecting tank. The runoff was collected in the collecting tanks after the rainfall events during 24 hours, at 8.30 a.m. (IST). The runoff in the tanks was thoroughly stirred and immediately a representative sample was collected in bottles of 1000 ml capacity. The sediment in the sample was measured after filtration and oven drying. Then the total sediment lost form each plot was estimated from the total tank capacity and the number of slots of the multi-slot divisor unit. All the runoff water collecting tanks were immediately emptied after the sample collection and were ready for the next rainfall event.

Data Analysis

The data analysis was made by using the statistical software SAS 9.2 (Statistical analysis software). The software needs the data in Julian date format. The Julian date started with 176 *i.e.* for 25th June, 2013 and ended with 243 *i.e* 31st August, 2013. This is the duration of the recording of observations during the considerable amount of rainfall in the study area. The data were analysed by using the GLM (General linear model) procedure that uses the method of least squares to fit general linear models. Fisher's Least Significance Difference (LSD) test was also used in analysis of variance (ANOVA) to create confidence intervals for all pair wise differences between factor level means while controlling the individual error rate to a significance level you specify. Fisher's LSD method then uses the individual error rate and number of comparisons to calculate the simultaneous confidence level for all confidence intervals. This simultaneous confidence level is the probability that all confidence intervals contain the true difference. The data constitutes 26 rainfall-runoff-sediment yield events, which were recorded over 24 hours rainfall events and their main characteristics and the corresponding hydraulic responses were evaluated.

Results and Discussion

Rainfall, Runoff and Soil Nutrient Loss

The maximum amount of rainfall occurs during the summer months from June to September. The winter months remain relatively dry. Daily rain fall in a rainy month shows considerable variation affecting the runoff and sediment production from the plots. The total quantity of rainfall received during the year 2013 was 3356.4mm. The maximum daily rainfall of 147mm was observed on 28, June, 2013. The maximum monthly rain 1093.9mm was received during the month of July 2013. Table 3 indicates medium to high content of organic matter during the cultivation of crops. The trend of the results of soil sample nutrient analysis showed that higher content of N, P and K in lower part of the runoff plots in comparison to their middle and uppers parts. This is due to the removal of nutrient from upper part to lower part of the land along with runoff water, which is corroborating the results obtained by previous studies (Sehgal and Abrol, 1944; CSWCRTI, 1994). The displaced nutrients are obstructed at the lower bunds of the runoff plots as shown in the Figures 9 and 10. The slope of the land plays an important role in accelerating the velocity of runoff water at higher intensity rainfall.

Table 3: Soil nutrient status of runoff plots.

Treatments		Specification			
		Nitrogen (kg/ha)	Phosporus, P_2O_5 (kg/ha)	Potassium (kg/ha)	Organic Carbon (g/kg)
T_1	U	263.4	86.3	156.8	9.0
	M	276.6	101.2	231.26	9.3
	L	376.3	158.6	78.33	13.6
T_2	U	225.7	74.7	201.42	10.9
	M	213.0	91.4	145.47	12.0
	L	313.2	95.6	111.9	16.3
T_3	U	250.8	65.7	44.76	10.3
	M	275.9	107.2	55.95	9.4
	L	295.9	110.9	74.6	11.5
T_4	U	250.8	89.3	67.14	10.4
	M	213.2	83.92	149.2	10.0
	L	285.8	91.7	208.88	9.5
T_5	U	238.3	72.3	264.83	17.2
	M	238.6	124.2	190.23	13.6
	L	351.2	91.1	197.69	17.2
T_6	U	288.5	77.3	59.68	13.6
	M	200.7	32.8	96.98	7.9
	L	326.1	92.6	130.55	18.3

Note: U-Upper part; M-Middle part; and L-Lower part of the runoff plot

Figure 9: Vigorous growth of maize towards the lower side of the runoff plot

Figure 10: Vigorous growth of cowpea towards the lower side of the runoff plot

One way ANOVA with rainfall as an independent and runoff as dependent variable was carried out by using the GLM module of SAS. The results are given in Tables 4 and 5. This indicates that there is a significant difference among the treatments in producing runoff from different rainfall events. An average runoff of 1631 litres was resulted from an area of 100 sq.m from the annual rainfall. To determine the way in which the treatments differ from each other in resulting runoff, Fisher's LSD Test was used (Hyeter, 1986). This showed LSD value of 570.04 litres at 5% level. The treatment T_4 resulted in highest amount of runoff (2198.4 litres), whereas, the treatment T_6 resulted in lowest runoff (1120.5 litres). This is due to the fact that the canopy of cowpea crop covers the soil surface effectively and does not allow the raindrop to directly hit the soil surface and obstructs the overland flow considerably. The low energetic rainwater gets more time to infiltrate and thereby produces low runoff (Wasi Ullah, 1972). However, the maize which is an erosion permitting crop cannot obstruct the overland flow and causes more runoff. Among the other treatments, the treatments T_5, T_2, T_1 and T_6 are at par in producing the runoff at different rainfall events. This indicates the effect of inter-cropping of maize and cow pea can reduce the runoff considerably in comparison to the maize alone as shown in the box plot Figure 11.

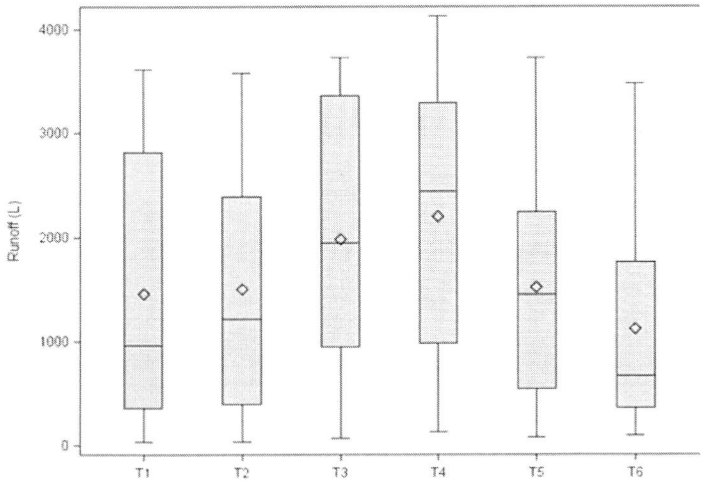

Figure 11: Box plot distribution of Runoff by Treatments

Table 4: Statistical analysis of Runoff (Litre) as a dependent variable

Fisher's Least Significance Difference Test

Alpha	0.05
Error Degrees of Freedom	6
Error Mean Square	705526.2
Critical Value of t	2.44691
Least Significant Difference	570.04

Means with the same letter are not significantly different

t Grouping	Mean	N	trt	
	A	2198.4	26	T4
	A			
B	A	1984.2	26	T3
B				
B	C	1516.2	26	T5
B	C			
B	C	1501.4	26	T2
B	C			
B	C	1459.7	26	T1
	C			
	C	1120.5	26	T6

Table 5: Statistical analysis of Sediment Yield (Grams) as a dependent variable

Fisher's Least Significance Difference Test

Alpha	0.05
Error Degrees of Freedom	6
Error Mean Square	553285
Critical Value of t	2.44691
Least Significant Difference	504.8

Means with the same letter are not significantly different.

t Grouping	Mean	N	trt	
	A	865.8	26	T4
	A			
B	A	620.7	26	T2
B	A			
B	A	555.7	26	T3
B	A			
B	A	466.5	26	T5
B				
B		356.5	26	T1
B				
B		289.5	26	T6

Similarly, the Fisher's LSD test by taking the sediment yield as a dependent and rainfall event as independent variable resulted highest sediment yield (865.8 g) for the treatment T_4 and the lowest for the treatment T_6 (289.5 g), which were significantly different from each other. The LSD value for the sediment yield was 504.8 g at 5% level. This might be due to the fact that cowpea as a cover crop does not allow the raindrops to hit the soil surface directly and displace the soil particles from its parent mass, which causes lower sediment yield. Reversely the maize crop canopy is unable to obstruct the raindrops from hitting the soil surface directly and thereby resulting in more sediment loss. The flow of runoff also aggravates the removal of sediment from the soil surface faster. As far as the other treatments in resulting sediment yield is concerned, the treatments T_4, T_2, T_3, T_5 responded similarly. Similarly, the treatments T_2, T_3, T_5, T_1, T_6 also responded at par. This might be due to the intense rainfall about 147 mm occurred on 28th June, 2013. High intensity rainfall can cause a complex hydrological environment causing unexpected sediment yield from a given cropping pattern. It can be said that the inter-cropping of maize along with cowpea can cause considerable lower sediment yield. This is given in Table 4 and as shown in the box plot Figure 12. Average sediment of 525.8 grams was resulted from the annual rainfall and from an area of 100 sq.m.

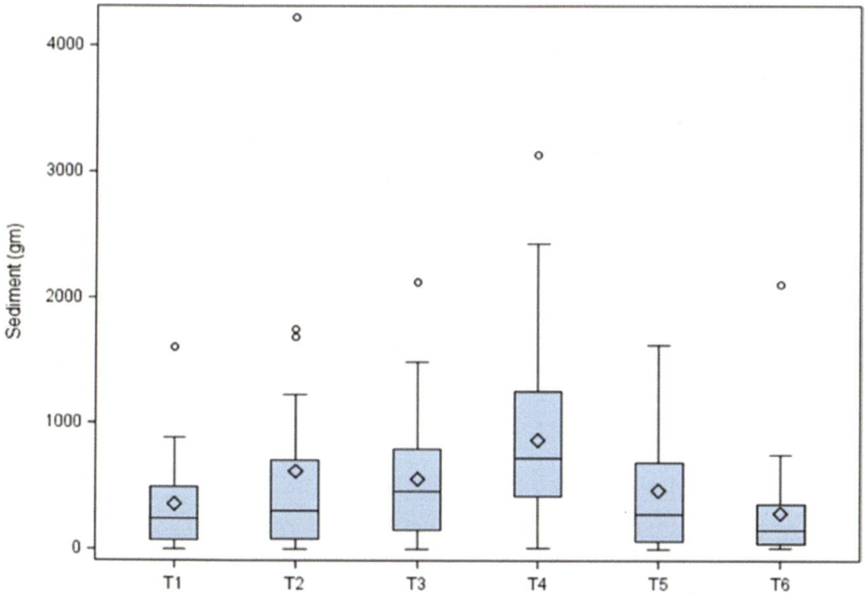

Figure 12: Box plot distribution of Sediment yield by Treatments

Higher value of co-efficient of determinations (R^2) for rainfall versus runoff (0.98) and rainfall versus sediment yield (0.94) indicate that the runoff and sediment yield data are highly correlated with rainfall data. Positive correlation exists between rainfall versus runoff and rainfall versus sediment yield. There occurs a moderate variation among the runoff data, whereas the sediment yield data are highly variable. For a given treatment, high rainfall causes higher value of runoff and higher value of sediment yield. This is in accordance to the concept that high intense rainfall is able to dislodge the soil particles from the parent mass and the higher runoff aggravates the removal of dislodged soil particles. The rainfall intensity of 5.6mm/hr was occurred on day 198, which was high in comparison to the day on 179, where the low rainfall intensity of 3.1mm/hr was received, as presented in the Figures 13 and 14.

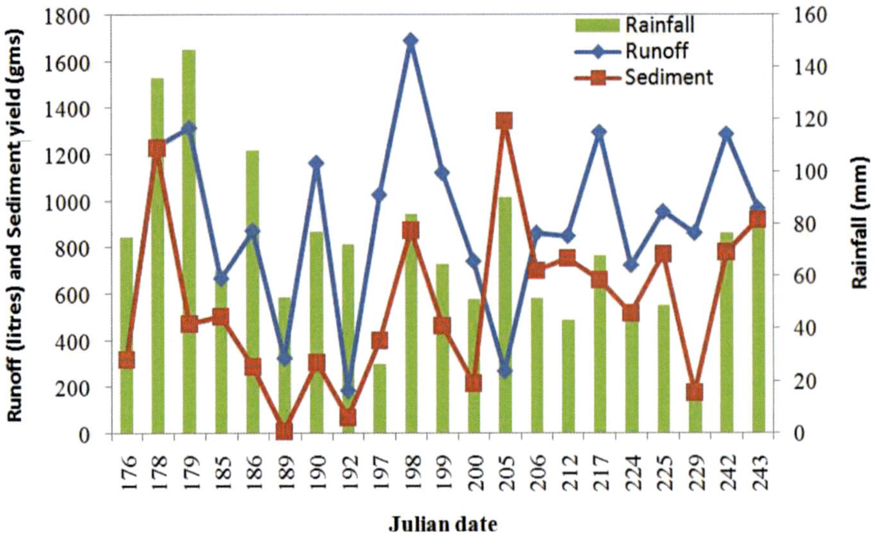

Figure 13: Variation of runoff and sediment yield from runoff plots at different rainfall events

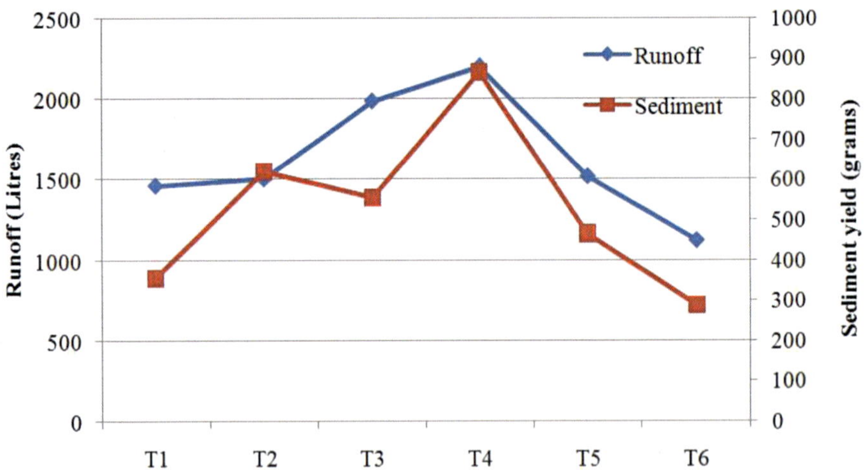

Figure 14: Variation of runoff and sediment yield at different reatment levels

Conclusions

The experiment conducted on runoff and sediment loss in the valley and foot-hill region of Arunachal Pradesh has generated valuable information of the considerable soil loss from the cultivated lands in mild sloping land. This is one of the major factors for the low productivity of the crops cultivated in Arunachal Pradesh, where major cultivated lands exist on sloping lands. The main cause of the considerable soil loss is high intense rainfall that persists for a longer period of about seven months of a year. The sandy and porous nature of soil also aggravates the situation. The farmers usually cultivate maize as one of the remunerative Kharif crop. This crop alone is not able to reduce soil erosion under this agro-climatic condition. The farmers can go for a inter-cropping of maize with cowpea which can reduce the soil erosion considerably in the slopping lands. It is found that maize-cowpea inter-cropping at 2:4 resulted in 38.6% less runoff and 30.7% less sediment loss. Hence, the farmers can be suggested not to go for single crop like maize instead to go for inter-cropping of maize and cowpea.

Acknowledgement

The authors are grateful to the Central Agricultural University, Imphal, Manipur to provide both financial assistance and laboratory facility to conduct the experiment at its College of Horticulture and Forestry, Pasighat, East Siang district of Arunachal Pradesh campus.

References

1. Central Soil and Water Conservation Research and Training Institute Annual Report: 1994-95, Dehradun, India: 195.
2. Hayter, A.J. 1986. The maximum family-wise error rate of Fisher's least significance difference test. Journal of the American Statistical Association, 81: 1000-1004.
3. http://www.fao.org/docrep/t0848e/t0848e-08.htm- Chapter 3: Field Plots
4. Jackson, M.L. 1973. Soil chemical analysis. Prientice Hall of India Pvt. Ltd., New Delhi.
5. Michaelides, K., Lister, D., Wainwright, J., & Parsons, A.J. 2009. Vegetation controls on small-scale runoff and erosion dynamics in a degrading dryland environment, Hydrological Processes, 23: 1617-1630.
6. Michael, A.M. 2009. Irrigation: Theory and Practice, Vikas Publishing House Pvt. Ltd, Noida, U.P.
7. Nearing, M.A. 2004. Soil Erosion and Conservation, In: Environmental Modelling: Finding Simplicity in Complexity, ISBN 100-471-49618-9, John Wiley & Sons Ltd.

8. Nett, Mary and Hendley Paul 2002. Designing Effective Runoff Research Studies: A Review of Issues of Scale, Pesticide Environmental Fate, American Chemical Society Symposium Series, Chapter 3, 813: 23–41.

9. Satapathy, K.K. 2003. Soil and Water Conservation Measures for Sustainable Hill Agriculture, Approaches for Increasing Agricultural Productivity in Hill and Mountain Ecosystem, ICAR Research for NEH Region, Umiam, Meghalaya, India: 03-25.

10. Sehgal, J. and Abrol, I.P. 1994. Soil degradation in India: Status and Impact. New Delhi: Oxford and IBH: 80.

11. Singh, A. and Singh, M.D. 1978. Effects of various stages of shifting cultivation or soil erosion from steep hill slopes, Indian For., 106: 115-121.

12. Wainwright, J., Parsons, A.J. & Abrahams, A.D. 2000. Plot-scale studies of vegetation, overland flow and erosion interactions: case studies of Arizona and New Mexico, Hydrological Processes, 14: 2921-2943.

13. WasiUllah, Gupta, S.K. and Dalal, S.S. 1972.Hydrological Measurements for Watershed Research, Jugal Kishore and Co., Rajpur Road, Dehradun, India.

❑❑❑

Chapter 10

Framework for Assessing Soil Security in Watersheds: A Case Study

Hrittick Biswas, A. Raizada, Suresh Kumar, B.S. Naik, A.S. Morade and P. Mohan Kumar

ICAR-Indian Institute of Soil and Water Conservation, Research Centre, Bellary-583 104, Karnataka, *E-mail: hritbis@yahoo.co.in

ABSTRACT

The last few decades have witnessed widespread land degradation due to the need to produce more from limited resources. With no future possibility of a let up in human-induced pressure, the challenge is to maintain the capacity of soils to continue providing economic goods and ecosystem services. Soil sustainability has been widely recognized as the most vital part of the solution to extent global environmental challenges. However, the value assigned to soils by society at large is not encouraging. Hence, the concept of soil security was recently developed to serve as a bridge and to inter-link the above-mentioned concerns. Soil security refers to the conservation and enhancement of soil resources for ensuring continuous food, fibre and fresh water production, energy and climate sustainability, maintenance of biodiversity, and the overall protection of ecosystem goods and services. Soil security requires that a value be placed on soil by all stakeholders, and that the value is supported by appropriate policy that ensures a secure soil by encouraging judicious soil management practices.

The unit often employed to systematically analyse the direct and indirect impacts of natural resource management (NRM) activities is a watershed. Watershed based approach not only involves the planning and implementation of integrated soil and water conservation measures from the ridge to the outlet, but also allows accurate monitoring of components of water balance or hydrologic cycle, sediment, energy, carbon and nutrients balances in a given eco-system. Therefore, watersheds offer appropriate ground for the study and assessment of soil security. We made an attempt to compare the soil security status of four different watersheds, typically representing the Central Dry Zone of Karnataka through a framework encompassing the five dimensions of soil security, *viz.* capability, condition, capital, connectivity and codification. Robust sets of indicators were developed for each dimension and assigned scales (0-1) and weightages as per their importance. The scores obtained for each indicator were aggregated to obtain a comprehensive soil security index (SSI) for the whole watershed. The SSI is expected to serve as a single and dynamic composite index for comparing different regions and prioritizing and drafting holistic plans for NRM for sustaining the flow of soil ecosystem services. The results of our assessment placed Gandasi and Bukkasagara watersheds ahead of Sukaligarahatti and Sugur watersheds in terms of SSI.

Keywords: Soil security index, watershed, natural resource management

Introduction

Continued human-induced pressure on land to meet the food and fibre demands has resulted in decline in essential soil functions. The present-day challenge is to maintain the capacity of soils to continue providing economic goods and ecosystem services. Although the present-day concerns of food, water, energy security and climate change abatement have 'soils' as the most vital component, the value assigned to soils by society at large is not encouraging (Bouma and McBratney, 2013). Therefore, a sustainable development concept for soil termed as 'soil security' referring to the conservation and enhancement of soil resources for ensuring continuous food, fibre and fresh water production, energy and climate sustainability, maintenance of biodiversity, and the overall protection of ecosystem goods and services (McBratney *et al.*, 2012), was coined. Long term soil conservation to maintain/enhance its 'functional capacity' requires deep understanding and involvement of all the stakeholders directly and indirectly dependent on soil. Soil security requires that a value be placed on soil by all stakeholders, and that the value is supported by appropriate policy that ensures a secure soil by encouraging judicious soil management practices. Soil security can be assessed through the integration of five dimensions, *viz.* condition, capability, capital, connectivity and codification (McBratney *et al.*, 2014). Comparison of soil security status among different regions or even for the same region on temporal scales requires quantification of the dimensions by employing appropriate and sensitive indicators.

Watersheds are units often employed to systematically analyze the direct and indirect impacts of natural resource management (NRM) activities by different agencies. Watershed based approach not only involves the planning and implementation of integrated soil and water conservation measures from the ridge to the outlet, but also allows accurate monitoring of components of water balance or hydrologic cycle, sediment, energy, carbon and nutrients balances in a given eco-system. Further, any watershed program ensures the active participation of Government agencies, project implementers and farmers (or beneficiaries) directly or indirectly dependent on the soils. Therefore, watersheds offer appropriate ground for the study and assessment of soil security. Against this background, an attempt was made to quantify soil security on watershed scales for the first time in India through a framework, involving the five dimensions of soil security. Easy to measure indicators selected with the help of expert opinion were employed to generate a single and dynamic composite soil security index.

Materials and Methods

Study Area

The Indian state of Karnataka has been divided into ten agro-climatic zones, based on the integration of data on rainfall, soils, climate and also the cropping patterns. These zones formed the basis for the identification of problems, formulation of research programmes and recommendations for adoption and implementation of development programmes for all land-based activities (Government of Karnataka, 2014). Four watersheds, typically representing the Central dry zone and from three districts of Karnataka, were selected for this assessment study (Figure 1). Brief descriptions of each of the watersheds are given below:

Sukaligarahatti watershed: This watershed, spanning 10.47 km^2 is situated in the Kadur *taluk* of Chikmagalur district and consists of five villages. The watershed receives an annual rainfall of about 686 mm, a bulk of it during the summer (*kharif*) season. The terrain in the upper reaches is gently sloping with an average slope of 2-5%. Soils are mainly red loamy Inceptisols and Alfisols. Rainfed agriculture is practised in 85% (890 ha) of the watershed area.

Sugur watershed: The Sugur watershed project is situated in the Hiriyur *taluk* of Chitradurga district in Karnataka. The Sugur watershed covers an area of 11.89 km^2 spread over four villages. The watershed has typical agrarian features of the semi-arid tropics – low rainfall (650 mm per annum), shallow red soils, small land owners (72%), poor level of education, traditional agricultural practices, high soil erosion rates, excessive groundwater extraction and poor recharge.

Bukkasagara watershed: The watershed is located in Kadur *taluk* of Chikmagalur district. The watershed (13°40′ to 13°50′ N latitude and 75°50′ to 75°57′ E longitude) covers a geographical area of 11.30 km^2. The area receives an annual rainfall of 600-650 mm, which is erratic and uncertain. More than 74% of the agriculture is carried out under rainfed conditions. The land is undulating with 8 to 15 % slope. The depth of soil ranges from 10 to 60 cm. Most of the land in the watershed falls under class III (180 ha), class IV (220 ha). Nearly 26 % of the land is under coconut plantation and seasonally irrigated crops. Major crops grown under rainfed conditions are potato, sunflower while commercial crops like maize and vegetables are cultivated under borewell irrigation.

Gandasi watershed: The watershed is located in the Arsikere *taluk* of Hassan district. The watershed (13°12′ to 13°15′ N latitude and 76°19′ to 76°20′ Long.) covers an area of 11.62 km^2, which supports 560 families that have settled in five villages. The annual rainfall is about 750 mm, which is bimodal – *kharif* rains in June-July and again in September-October. The general land slope is undulating with slopes ranging from 3-10%. The soils are Entisols with variable depth, which ranges from 7.5 cm in the upper reaches to about 90 cm in the valley portions. As a consequence, soil and water erosion is widespread and there is very little moisture retention in arable lands, where no conservation measures exist. Nearly 70% of the arable land is classified as Class III, 20% as Class II and 20% as Class IV of land capability classification. Under rainfed conditions *kharif* crops like green gram, sesame, potato and sunflower are grown. Under irrigated conditions, commercial crops like maize and vegetables are grown.

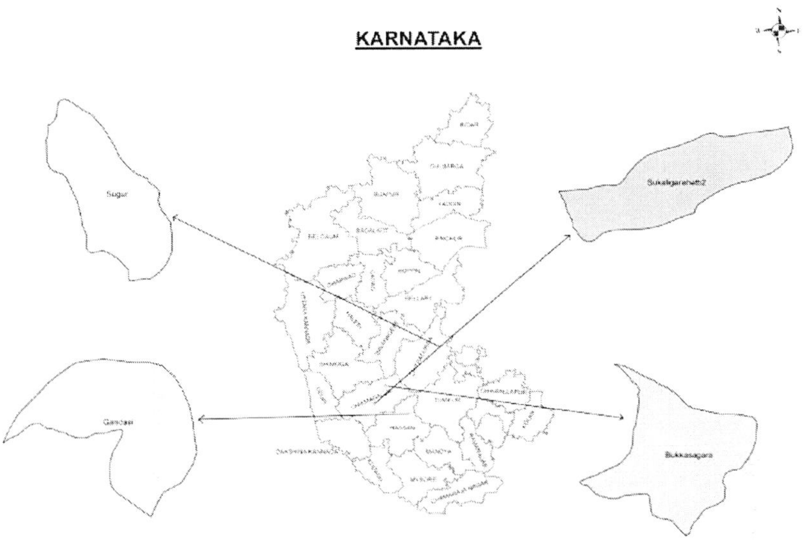

Figure 1: The study area

Selection of Indicators

Indicators are statistics or measures that relate to a condition, change of quality, or change in state of something valued. They provide information and describe the state of the phenomena of interest, but with a significance beyond that directly associated with an individual parameter (OECD, 1993). Indicators need to be developed according to their perceived applications, and this requires reliable statistics and raw data. Because of regional requirements and priorities, global agreement on a single set of indicators would be difficult and unnecessary for many issues.

The selection of the 'right' indicators to best describe the status of each of the dimensions can be debated ad nauseam (Fredericks, 2012) and can vary due to the choice of the researchers', availability of data or ease of measurement. The following subsections are devoted to the selection of indicators that reflect the status of five dimensions in the watersheds. These indicators were combined based on certain criteria to construct the dimensional indices, for example, the condition index or capability index and finally the soil security index (SSI). It is worth mentioning here that the indicators were selected based on their ability to define the dimension as closely and as quantitatively as possible. The suggested list of indicators has been provided in Table 1:

Capability: Capability of soil is usually defined as the optimal capacity of the soil in the reference state. The properties of a pristine or undisturbed soil in a watershed can be considered as the reference, wherein the soil functions at its optimal level. The indicators selected for quantifying the capability dimension are pH, soil organic carbon, clay content, bulk density and soil erodibility factor.

Condition: Soil condition refers to the present capacity or state of the soil after it has been subject to cultivation practices, erosion, fertilization, conservation measures, etc. The indicators chosen for soil condition are more or less the same as those for the capability dimension, with some additional indicators, such as per cent land cover, soil loss, additional area under salinity/alkalinity, etc.

Capital: Soil capital is usually considered from a monetary point of view. The closest indicator of soil capital is crop productivity. Other indicators are resource use efficiency and resource savings.

Connectivity: Connectivity emphasizes on the value attached by a farmer or society (stakeholders in a watershed) to soil. The indicators selected for connectivity are fertilizer and/or compost application rates/frequencies (based on sound soil testing), choice of crops and cropping intensity, method and frequency of irrigation, and soil conservation/moisture conservation measures adopted.

Codification: Codification refers to the importance given by the Government or implementing agencies to the soils of the watershed. The indicators chosen for this dimension are: area covered under soil and water conservation programmes, soil health cards distributed, subsidies on inputs, such as fertilizers, composting units, number of micro-irrigation units, number of *krishi vigyan kendras* or subject matter specialists per km^2.

Table 1: Suggested list of indicators for computation of soil security index

S. No.	Capability (C_1)	Condition (C_2)	Capital (C_3)	Connectivity (C_4)	Codification (C_5)
1.	pH	Soil loss	Crop productivity	Soil-test based fertilizer application	Area under soil and water conservation measures
2.	Soil organic carbon	Salinity/alkalinity	Resource saving	Application of organic manures	Fertilizer distribution
3.	Clay content (%)	Land use/land cover	Resource use efficiency	Cropping intensity	Subsidies
4.	Bulk density	Soil permeability		Irrigation	Soil health cards
5.	Soil erodibility factor			Tillage	No. of SMS/KVK per ha
6.	Soil respiration				
7.	Land slope (%)				

Normalization of indicators

Once the indicators are selected, they have to be normalized so that they can be compared. Since indicators are measured in different units and scales, we adopted the methodology used in UNDP's Human Development Index (UNDP, 2006) to normalise the indicators. That is, in order to obtain figures that are free from the units and also to standardise their values, first they were normalised so that they all lie between 0 and 1. Before doing this, it is important to identify the functional relationship between the indicators and sustainability. Two types of functional relationship are possible: (a) positive relationship: the function in question increases with increase in the value of the indicator, (b) negative relationship: when the function decreases with increase in the value of indicator.

If Z represents the ith indicator of the jth dimension (i = 1,2,....n and j = 1ij2,..5), the positive relationships can be expressed as equation 1, and the negative, as equation 2.

$$Zij = \frac{Xij - Min\{Xij\}}{Max\{Xij\} - Min\{Xij\}} \qquad -(1)$$

$$Zij = \frac{Max\{Xij\} - Xij}{Max\{Xij\} - Min\{Xij\}} \qquad -(2)$$

Aggregation: For the dimension C_j,

$$Cj = \sum_{i-1}^{n} Xij \times Wij$$

Soil Security Index: If equal weights are assigned to all the dimensions, soli security index can be expressed as:

$$SSI = \sum_{j=1}^{5} Cj$$

Soil samples: Thirty samples from each watershed were collected at 500 m X 500 m grids and analyzed for physico-chemical properties.

Socio economic survey: 75 farmers in each watershed (marginal, small, medium and large) were interviwed through structured schedules.

Results and Discussion

Table 2 shows the computed indices for all the five soil security dimensions and the overall SSI. The watersheds placed in terms of the SSI are: Gandasi H" Bukkasagara>Sukaligagahatti>Sugur. A higher SSI indicates that both farmers and the district functionaries related to the watershed have a better understanding about the value of soil sustainability. It may be noted that all the watersheds represent the same agro-climatic region and have similar conditions of soil development. Therefore, all the watersheds are almost similar in the capability index, implying that the initial soil properties of the undisturbed soils are almost similar. The difference is created, when the farmers or residents start using the soils for satisfying their food, fuel, fibre and fodder requirements. Thus, it can be seen that the soils of Sugur watershed have been deteriorated more compared to the other three watersheds. This has also been reflected in the lower crop yields of Sugur watershed, thereby resulting in the lower soil capital index. Most of the farmers of Sugur and Sukaligarahatti are found wanting in caring for their soils. This is evident from the higher proportion of land left fallow or put solely to commercial cultivation without undertaking adequate soil and water conservation measures. So, the farmers are less connected to their soils as reflected in the lower connectivity index. Excepting Sugur, the other three watersheds are similarly placed in terms of codification index, which also could be improved by pro-active involvement of Government agencies. However, it seems that for Sugur watershed, linkages between the beneficiaries and Government through the implementing agencies have been far from satisfactory. On the contrary, the extension workers in the other three watersheds are more pro-active as compared to those in Sugur watershed, resulting in the higher values of codification index. Among the farmer groups, small and marginal farmers performed better than medium and large farmers of both watersheds in terms of SSI. This may be attributed to the fact that marginal farmers do not have ready access to and are poor managers of resources, while the small and medium farmers judiciously utilize available resources. Further, the large farmers mostly have diversified businesses, reside in big towns or cities and either leave their land fallow or lease out their land to landless and marginal farmers.

Table 2: Computed soil security indices for the four watersheds.

Farmer category	Capability (C_1)	Condition (C_2)	Capital (C_3)	Connectivity (C_4)	Codification (C_5)	Soil Security Index
			Bukkasagara			
Marginal	0.59	0.57	0.6	0.68	0.58	0.604
Small	0.6	0.66	0.72	0.72	0.66	0.672
Medium	0.62	0.65	0.75	0.72	0.66	0.680
Large	0.63	0.67	0.68	0.68	0.66	0.664
Overall	*0.61*	*0.64*	*0.69*	*0.70*	*0.64*	0.655
			Gandasi			
Marginal	0.6	0.56	0.6	0.65	0.54	0.590
Small	0.62	0.68	0.7	0.69	0.64	0.666
Medium	0.62	0.7	0.72	0.73	0.64	0.682
Large	0.64	0.65	0.75	0.7	0.68	0.684
Overall	*0.62*	*0.65*	*0.69*	*0.69*	*0.63*	0.656
			Sugur			
Marginal	0.58	0.51	0.55	0.52	0.48	0.528
Small	0.59	0.56	0.62	0.61	0.5	0.576
Medium	0.61	0.58	0.65	0.63	0.54	0.602
Large	0.62	0.63	0.6	0.62	0.6	0.614
Overall	*0.60*	*0.57*	*0.61*	*0.60*	*0.53*	0.580
			Sukaligarahatti			
Marginal	0.6	0.55	0.6	0.58	0.55	0.576
Small	0.62	0.6	0.62	0.63	0.61	0.616
Medium	0.61	0.65	0.66	0.65	0.62	0.638
Large	0.63	0.65	0.65	0.65	0.65	0.646
Overall	*0.62*	*0.61*	*0.63*	*0.63*	*0.61*	0.619

Conclusion

Soil security status of Gandasi and Bukkasagara watersheds was better than Sugur watershed. The Soil Security Index can, therefore, be gainfully used not only to compare soils of different regions, but also to serve as a guide to state level functionaries to prioritize and draft holistic plans for sustaining the flow of soil ecosystem services. SSI is a tool to answer the question "How much do we care for our precious soils?".

References

Bouma, J. and McBratney, A.B., 2013. Framing soils as an actor when dealing with wicked environmental problems. *Geoderma,* **200-201**: 130-139.

Fredericks SE. 2012. Justice in sustainability indicators and indexes. *International Journal of Sustainable Development and World Ecology,* **19**: 490–499.

McBratney, A.B., Minasny, B., Wheeler, I., Malone, B.P., 2012. Frameworks for digital soil assessment. In: Minasny, B., Malone, B.P., McBratney, A.B. (Eds.), *Digital Soil Assessment and Beyond*. Taylor & Francis Group, London, pp. 9–14.

McBratney, A.B, Field, D.J. and Koch, A., 2014. The dimensions of soil security. *Geoderma*, 213, pp. 203-213.

❑❑❑

Chapter 11

Hidden Nutrient Cost: The Neglected Nutrient Budget for Carbon Sequestration

Trisha Roy[a*], A.K. Biswas[b], Pramod Jha[b] and Justin George[c]

[a]ICAR- Indian Institute of Soil and Water Conservation, Dehradun, [b] ICAR-Indian Institute of Soil Science, Nabibagh, Bhopal, MP, [c] ISRO-Indian Institute of Remote Sensing, Kalidas Road, Dehradun, trisha17.24@gmail.com

ABSTRACT

Approximately, half of all soil carbon in managed ecosystems has been lost to the atmosphere during the past two centuries due to cultivation and this loss represents a huge potential of the soil acting as a major carbon "sink" to store the atmospheric CO_2. Hence, addition of soil organic matter (SOM) and sequestration of C under such context is a complete win-win situation, as it would reduce the concentration of atmospheric CO_2 as well as result in an increment in the soil organic carbon (SOC) level and improve soil quality. This C increment would also increase soil aggregation, reduce erosion and arrest land degradation. However, for conversion of added C into stable humified fraction not only C but other building blocks like N, P, S, and several micro-nutrients are also required. It is estimated that for sequestration of 10,000 kg of C in stable humus, 833 kg of N, 200 kg of P and 143 kg of S are needed assuming C:N ratio of 12:1, C:P ratio of 50:1 and C:S ratio of 70:1. Thus, lack of nutrients could not only limit the plant biomass production but also limit the amount of C assimilated in the stable fraction of SOM in the absence of additional nutrients required over and above crop growth. This additional nutrient cost incurred for humification of added C in the stable fraction is neglected most of the time while adopting agricultural practices, such as conservation agriculture and other resource conservation technologies to improve soil health and mitigate climate change. This paper mainly reflects the importance of the N, P and S fractions in sequestering C, which should be taken into account for enhancing the potential of soil as a C-sink.

Keywords: Carbon sequestration, nutrient budget, CNPS ratio, C mineralization.

Introduction

The fifth assessment report of the Inter-governmental Panel on Climate Change (IPCC) and the recent Paris Convention held in 2015 aims to curb the green house gas emission by 40-70% in the next 30 years by 2050, so as to restrict the increase in the global mean surface temperature by 2°C (Metz *et al.,* 2005). Though the statement sounds simple and promising but the means to achieve this task at a global level is highly challenging. And if proper measures are not taken to achieve this, it will result in a catastrophe with the global mean surface temperature increasing by 3.5 to 4.8°C, and consequently, causing the warming up of the oceans, rise in the mean sea level (msl), shrinking of the glacier volume, flooding of the coastal areas, and not to forget a huge impact on the agricultural production scenario, with reduced crop productivity of major cereals like wheat, rice, maize and soyabean. Thus, limiting the warming process would require substantial technological, economic and institutional interventions, which in turn implies more efficient use of energy, greater use of low-carbon and no-carbon energy, improved carbon sinks and changes in the lifestyle and behavioural pattern of the global population.

The global carbon cycle distributes C mainly in three different pools: the atmosphere, the hydrosphere and the terrestrial ecosystem. The terrestrial ecosystem stores around 2000 Pg C, of which majority of the share (1500 Pg) is cached in the soil system amounting to three times as much C in the terrestrial biomass and as much as twice the C present in the atmosphere (Post *et al.,* 1982; Schlesinger, 1992; Eswaran *et al.,* 2000). The soil which acts both as a sink and source of CO_2 could be an important storage strategy through carbon sequestration and hoarding the C in soil over long term period to mitigate the impact of climate change (Lal, 2004 and 2006). The storage of C in soils, particularly by targeting degraded lands could help in increasing the size of the SOC pool, improve the quality of soil and also alleviate the effects of rising CO_2 concentration (Kirkby *et al.,* 2014). Thus, C sequestration through hoarding C in the soil system is a win-win situation in all scenarios with beneficial impact on the entire ecosystem as a whole. This also opens an avenue for an entire carbon sink which could help in mitigating the climate change problem.

However, addition of organic inputs to any land-use system does not lead to the linear increase in the organic C content. Many times researchers have reported very insignificant increase in the SOC despite addition of residues over long term. For instance, Soon (1998) reported no increment in SOC levels after 10 years of organic input addition to soil. Campbell *et al.* (1990) compared 31 years old experiment and observed no significant changes in the SOC levels in plots with and without residue addition. Thus, in addition to organic input there are certain things which need to be considered for the C sequestration. Himes (1998) established that the stable organic matter fraction for soils have almost constant C: N: P: S ratio. This ratio of 10000: 833: 200: 143 was used by Lal (2008) and verified for a large number of soil samples worldwide by Kirkby *et al.* (2011). This ratio makes it clear that for carbon to be humified and formed into stable humus fraction, besides organic input nutrients like N, P and S in the correct proportion would have to be supplied. This indicates that supplemental dose of the inorganic nutrients added with C residue would definitely impact the transformation of added organic matter to stable humus *i.e.* enhance the process of humification.

However, the information related to the impact of different supplemental nutrient dose on added crop residue is not widely available. This want of information triggered the formulation of the present experiment to evaluate the impact of different dose of N, P and S on the carbon mineralization from added crop residue. A laboratory incubation experiment was conducted under controlled condition and different doses of N, P and S was applied, whose impact on C release from wheat straw added to soil was studied.

Materials and Method

Collection of wheat straw

Wheat straw (*Triticum aestivum L.*) was collected from the ongoing field experiment at ICAR-Indian Institute of Soil Sciences, Bhopal, Madhya Pradesh and oven dried at 60°C and stored. The straw was cut into 5 mm pieces before uniformly mixing with the soil. A subsample portion was used to determine the total C, N, P and S content of the straw. The added wheat straw contained 42% C, 0.5% N, 0.022% P and 0.051% S, which is used for calculating the subsequent nutrient doses.

Collection of soil sample

Bulk surface (0-15 cm) soil samples were collected from ICAR-IISS, Bhopal, air-dried, ground and sieved through 2 mm sieve. The soil was clay loam in texture with 57.3 % clay, 26.8 % silt and 15.9 % sand. The total C content was 1.157 % analyzed using a TOC analyzer.

Experimental setup

Soil sample (100g) was incubated in 250 mL Borosil Erlymner conical flasks with and without wheat straw. For each treatment, there were three replicates. The conical flasks were fitted with rubber corks. To each rubber cork a 10 mL vial for NaOH trap was attached with the help of string. Wheat straw was mixed with soil @ 5 ton ha^{-1}, which was calculated to be 5 g straw per 100 g soil. This dose was finalized considering the optimum dose of residue or organics addition in different agro-ecological regions across the country. There were four nutrient treatments with added wheat residue (WR) (T_1) soil + WR + no nutrient solution, (T_2) soil + WR + nutrient solution in optimum dose, (T_3) soil + WR + 2 × optimum nutrient dose, (T_4) soil + WR + 4 × optimum nutrient dose. Another set of treatment without residue addition (WOR) with same nutrient doses *i.e.* (T_5) soil + no nutrient solution, (T_6) soil + nutrient solution in optimum dose, (T_7) soil + 2 × optimum nutrient dose, (T_8) soil + 4 × optimum nutrient dose were also maintained. The added wheat straw and nutrient solution was mixed thoroughly with the soil. All the flasks were kept at room temperature and pressure at 70% field capacity. The moisture content was maintained throughout the experimental cycle by adjusting the weight loss in soil by addition of de-ionized water over the time period. The alkali trap method was used for determining the amount of C mineralized. The C mineralized as CO_2 was measured using NaOH trap method at different intervals of 2, 4, 8, 14, 25, 35, 51 and 100 days and excess NaOH was back titrated using 0.5 N HCl with saturated $BaCl_2$ and phenolphthalein indicator (Alef and Nannipipieri, 1995). After each titration, the vials were filled with fresh NaOH and wrapped with para-film to prevent any loss of evolved CO_2. The HCl used was standardized using 0.5 N sodium carbonate using methyl red indicator to know the exact strength of acid. For each set, replicate of blank sample with 100 mL deionized water was used to measure the atmospheric CO_2 contribution.

The amount of C evolved was calculated using the formula:

$$C\ \text{evolved(mg per 100 g soil)} = (B\text{-}S) \times N \times 6 \times (\text{equivalent weight of C})$$

Where B is the amount of HCl consumed in blank sample

S is the amount of HCl consumed in each sample

N is the normality of HCl

Calculation of Nutrient Dose

Considering the C:N:P:S ratio of 10000:833:200:143, the supplementary nutrient dose was calculated to allow humification of 18 % of the C added through wheat straw. Under such circumstances, the contribution of the soil nutrient pool is nil, P present in the straw becomes the limiting nutrient element for conversion of C to stable humus pool. The additional amount of N, P and S (0.65 mg, 0.65 mg and 0.28 mg, respectively per 100 g soil) was added through nutrient solution. Solution containing ammonium sulphate and potassium di-hydrogen phosphate was used for getting the required nutrient concentration per 5 mL of solution which was added to soil and mixed homogeneously. In this study, the contribution from inorganic nutrient source was disregarded as their concentration is negligible in terms of absolute quantity compared to the added nutrients (Kirkby *et al.*, 2013).

Results and Discussion

Carbon efflux (Vertisol)

Results from the 100 day incubation experiment showed that in case of Vertisol, there was significant increment in C mineralization due to addition of wheat straw. The soil without any straw addition showed very low rate of C mineralization. After the completion of 100 days incubation, the cumulative C mineralization was 7.2 to 7.6 percent of the soil C, where no residue addition was there. However, in Vertisol with residue addition the cumulative C mineralization ranged from 14.4 to 20.6 per cent. The impact of supplementary nutrient addition was prominent where the highest efflux was obtained, where four times the optimum nutrient dose had been applied and the minimum CO_2 efflux was observed, where no nutrient had been added. But the nutrient dose did not enhance the CO_2 efflux from soils without any residue amendment. This indicates that after removal of the coarse fraction SOM through sieving (<0.5 mm) the soil did not contain much labile fraction of soil organic C. The absence of the labile pool did not allow the microbes to proliferate despite the abundant supply of N, P and S through nutrient solution. But availability of the labile pool immediately doubled the C mineralization rate (14 %), even when nutrient was in short supply. Hence, we could say that the labile C availability was more prominent in controlling the microbial activity rather than the nutrient supply. However, comparing the three nutrient doses with the additional C availability, it is noticed that microbial proliferation is increased rapidly as reflected by the enhanced CO_2 efflux rates (17.1 to 20.6 %).

When the mineralization of straw C was considered exclusively [(C efflux from soil+ straw) - (C efflux from soil) = C efflux from straw] 88.3 % of the straw C was mineralized within 100 days period, when quadruple dose of supplementary nutrient was added. The range of C mineralization was from 52.2 to 88.3 % with the minimum C efflux from soil without any supplementary nutrient dose (Fig 1). The CO_2 efflux almost flattened at 51 days after incubation indicating that majority of the labile C in the straw had been dissimilated by the micro-flora by this period. Similar trend was observed for the soil without the straw addition, though there was a gap between the amount of C evolved with and without straw addition.

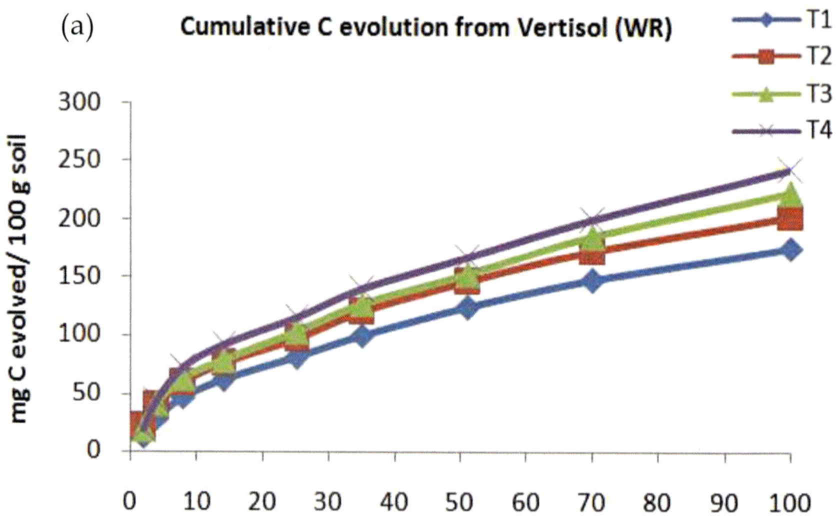

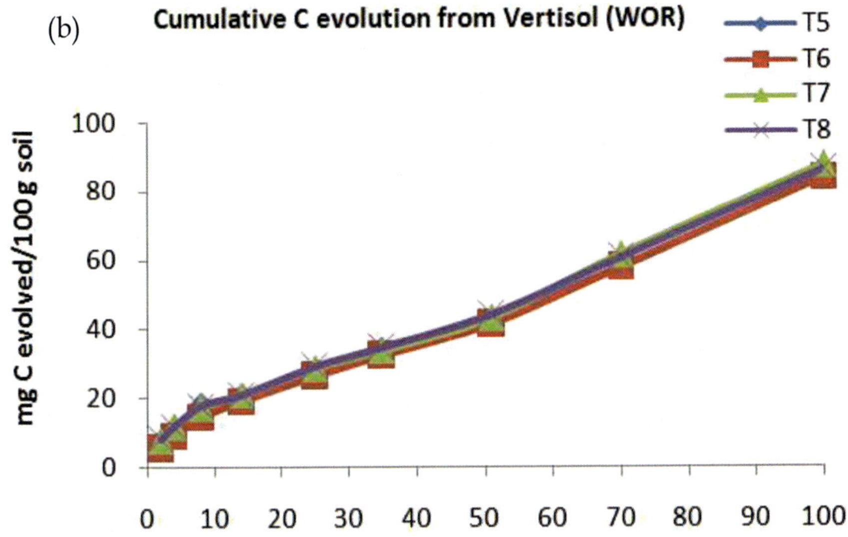

Fig. 1 a,b. Cumulative evolution of C from Vertisol with residue (WR) amendment and without residue (WOR) amendment

The soil C is more stable fraction compared to the freshly added wheat straw, thus, it is less likely to be impacted by the additional supply of nutrient. The freshly added plant residues or residue at various stages of decomposition are referred to as the light fraction of SOM, which is most labile and responsive to various alterations in the management practices (Six *et al.*, 2002; Solomon *et al.*, 2000; Janzen *et al.*, 1992). Thus, nutrient addition has more impact on release of C from the residue amended soils than the soils with no available labile fractions in the form of residue. Also, with increased nutrient dose, there is higher C release but optimum nutrient dose results in mineralization of 57% of the added C after which it is stabilized. Thus, addition of nutrients results in higher mineralization rates and better C sequestration.

Conclusions

This study indicates the impact of supplemental N, P and S doses on the behaviour of inherent soil C and also the C added through wheat straw. The three soils although followed similar trends but the extent of C mineralization gives clear cut impact on the management strategies to be adopted in different soil types. The nutrient addition following the CNPS of stable organic matter and depending on the type of residue or organic matter added will have an immense impact on the amount of C being converted to the stable organic fraction. If excess nutrient addition is practised, this could lead to more dissimilation of C from added residue rather than conversion into stable pools. This implies that only residue addition without the proper supply of nutrient would not help in C sequestration in soils. Rather it would deplete the soils of N, P and S, which would be required for synthesis of stable organic matter fraction and ultimately lead to lower net primary productivity of crops.

References

Alef, K. and Nannipipieri, P. 1995. Methods in Applied Soil Microbiology and Biochemistry. Academic Press, London.

Campbell, C.A., Bowren, K.E., Schnitzer, M., Zenter, R.P. and Townley-Smith, L. 1991. Effect of crop rotations and fertilization on soil organic matter and some biochemical properties of a thick Black Chernozem. *Canadian Journal of Soil Science,* 71: 377–387.

Eswaran, H., Reich, P.F. and Kimble, J.M. 2000. Global carbon stocks. In: Lal, R., *et al.* (Eds.), Global Climate Change and Pedogenic Carbonates. CRC Press, Boca Raton, FL, pp. 15-25.

Himes, F.L. 1998. Nitrogen, sulphur and phosphorus and the sequestration of carbon. *Soil Processes and the Carbon Cycle.* CRC Press Boca Raton, FL, pp 315-319.

Janzen, H. H., Campbell, C. A., Brandt, S. A., Lafond, G. P. and Townley-Smith, L. 1992. Light fraction organic matter in soils from longterm crop rotations. *Soil Science Society of America Journal,* 56: 1799–1806.

Kirkby, C.A., Kirkegaard, J.A., Richardson, A.E., Wade, L.J., Blanchard, C. and Batten, G. 2011. Stable soil organic matter: a comparison of C: N: P: S ratios in Australian and other world soils. *Geoderma,* 163(3): 197-208.

Kirkby, C.A., Richardson, A.E., Wade, L.J., Batten, G.D., Blanchard, C. and Kirkegaard, J.A. 2013. Carbon-nutrient stoichiometry to increase soil carbon sequestration. *Soil Biology and Biochemistry,* 60: 77-86.

Kirkby, C.A., Richardson, A.E., Wade, L.J., Passioura, J.B., Batten, G.D., Blanchard, C. and Kirkegaard, J.A. 2014. Nutrient availability limits carbon sequestration in arable soils. *Soil Biology and Biochemistry,* 68: 402-409.

Lal, R. 2004. Soil carbon sequestration impacts on global climate change and food security. *Science,* 304(5677): 1623-1627.

Lal, R. 2006. Enhancing crop yields in the developing countries through restoration of the soil organic carbon pool in agricultural lands. *Land Degradation & Development,* 17(2): 197-209.

Lal, R. 2008. Carbon sequestration. *Philosophical Transactions of the Royal Society B: Biological Sciences,* 363(1492): 815-830.

Metz, B., Davidson, O., De Coninck, H.C., Loos, M. and Meyer, L.A.2005. IPCC, 2005: IPCC special report on carbon dioxide capture and storage. Prepared by Working Group III of the Inter-governmental Panel on Climate Change. *Cambridge, United Kingdom and New York, NY, USA,* 442.

Post, W.M., Emmanuel, W.R., Zinke, P.J. and Stangenberger, A.G. 1982. Soil carbon pools and world life zones. *Nature,* 298: 156-159.

Schlesinger, W.H. 1995. An overview of the carbon cycle. In: Lal, R., Kimble, John, Levine, Elissa and Stewart, B.A. (eds). Soils and Global Change. Lewis Publishers, Boca Raton, FL, pp. 9-25.

Six, J., Conant, R.T., Paul, E.A., and Paustian, K. 2002. Stabilization mechanisms of soil organic matter: implications for C-saturation of soils. *Plant and Soil,* 241(2): 155-176.

Solomon, D., Lehmann, J. and Zech, W. 2000. Land-use effects on soil organic matter properties of chromic Luvisols in semi-arid northern Tanzania: carbon, nitrogen, lignin and carbohydrates. *Agriculture Ecosystems Environment,* 78. 203–213.

Soon, Y.K. 1998. Crop residue and fertilizer management effects on some biological and chemical properties of a Dark Grey Solod. *Canadian Journal of Soil Science,* 78: 707–713.

❏❏❏

Chapter 12

Soil Biochemical Properties and Water Infiltration Behaviour Under High Density Mango Plantation in Red Soils of Karnataka

A.S. Morade[1], M. Pannirselvam[2], R.M. Kurian[3], A.N. Ganeshamurthy[3] and H. Biswas[1]

[1]ICAR- Indian Institute of Soil and Water Conservation, Research Centre, Bellary., [2]ICAR- National Rice Research Institute, Cuttack, [3]ICAR-Indian Institute of Horticultural Research, Bangaluru

ABSTRACT

High density planting (HDP) is an intensive system of fruit cultivation which involves planting at lower spacing, maintenance of small tree canopy by adopting dwarfing rootstocks, growth retardants, training and pruning practices, which increases orchard efficiency. Present study was conducted in 15 years old established mango orchard in red soils under rainfed conditions at ICAR- Indian Institute of Horticultural Research, Bangalore to investigate the impact of two rootstocks (R), three spacing levels (S) and paclobutrozol doses (P) on soil biochemical properties *viz.* respiration rate, glomalin protein content, dehydrogenase activity, phosphatase activity and water infiltration rate. Paclobutrozol application (g of active ingredient/ year age of tree) was started from the fourth orchard year and its dose was stabilized by tenth orchard year. Therefore, 18 treatment combinations employed in split-split plot design. Alphonso variety grafted on unspecified seedling rootstock planted at 10 x 10 m without any paclobutrozol application served as an additional control treatment. A total of 19 treatments were replicated twice at same site. Statistical analysis revealed that soil biochemical properties were not altered by rootstocks, spacings as well as age-based paclobutrozol doses. HDP increased soil water infiltration by 46% compared to control (10×10 m). Both the rootstocks Vellaikolumban and Oluor had non-significant impact on water infiltration rates but it increased significantly with plant population. Infiltration rate of 100 mm hr^{-1},

79 mm $hr^{-1}$and 61 mm hr^{-1} were observed at spacing levels of 3×3 m, 5×5m and 7.5×7.5 m, respectively. Higher root density increased pore space not only in rhizosphere but also in subsurface soils in tree alleys which allowed higher water percolation in soil. The study concluded that HDP along with age-based paclobutrazol application in mango has not affected soil microbial activity but significantly improves soil water infiltration rate thereby retained surface run-off and minimized soil erosion.

Keywords: Age-based paclobutrazol application; high density mango; soil biochemical properties and soil water infiltration.

Introduction

Mango is major fruit crop of the tropics and subtropics and it is cultivated in at least 87 countries. Mango shares 33.8 per cent and 20.8 per cent of total fruits area and production, respectively. The average productivity of mango in the India is 8.35 t ha^{-1} (NHB, 2016). Poor mango yields of 4 to 9 t ha^{-1} is partly attributable to wide tree spacing, which are traditionally based on expected eventual tree size. Large tree size of mango creates a major problem in orchard management and contributes to lower orchard efficiency. The high-density orchard provides 8 to 9 times higher yields than the traditional densities. Hybrid mango variety Amrapali has been recommended by IARI to be grown at 2.5×2.5 m (triangular method) and accommodating 1,600 plants/ha as against 80 to 100 plants/ha under traditional planting density yielded upto 22 t ha^{-1}. Dashehari mango at 2.5×3 m (1,333 plants/ha) can also be raised under HDP with pruning and dehorning after the harvesting after 10th year of planting.

The control of tree growth retardants provides an attractive proposition in the absence of suitable dwarfing rootstocks for mango. The 2S and 3S enantiomer of paclobutrazol is acknowledged to be a specific inhibitor of gibberellins biosynthesis (Rademacher, 1995), with reduced internode extension as the main morphological effect.The paclobutrazol addition to soils of mango orchards in Petrolina, affected negatively the soil microbial community (Silva *et al.*, 2003). The average values for total number of bacteria, fungi and actinomycetes were reduced by 58, 28 and 28 per cent, respectively, compared to the paclobutrazol unamended soil. For the greenhouse experiment, the paclobutrazol application in the soils from Petrolina influenced negatively the dehydrogenase activity and the living hyphal length, but not the microbial biomass carbon.

Kotur (2006) observed significant increase in the root activity towards the trunk and close to soil surface and sparser root activity in the subsoil zone and in dripline area in paclobutrazol treated plants of Alphonso and reported decline in the root activity. Soil application of paclobutrazol (2.0 to 8.0 g a.i.) for two consecutive years decreased organic carbon. Whereas, phosphorus, potassium and calcium level increased at lower doses but decreased at higher dose. Calcium level was also decreased at higher doses. A similar trend was recorded in soil microbial level (Singh *et al.*, 2005). The findings indicate inhibitory effect of paclobutrazol at higher concentration (6 to 8 g a.i.) on soil nutrient status and microbial population.

There is little published information on soil microbial dynamics in relation to HDP of tropical fruit crops. However, there are evidences to suggest that the HDP could affect the soil temperature, root density, soil moisture regime and weed population. On this background, present study was conducted in established high density mango orchard with the objective to investigate impact of rootstocks, spacing levels and paclobutrazol doses on soil biochemical *viz.* respiration rate, glomalin protein content, dehydrogenase activity, phosphatase activity and soil physical property i.e. water infiltration rate.

There are ample indications on the role of planting density of tropical fruit trees on the soil properties through their influence on soil temperature, soil moisture, root density and weed infestation. There could be addition of more organic residues to the soil through leaf litter under high density planting, but at the same time there could be increased stress on the soil nutrient pool by the higher plant population. Hence, there is need to study these complexities and their interactions in relation to substrate dynamics and sustainability of high-density agro-systems in depth.

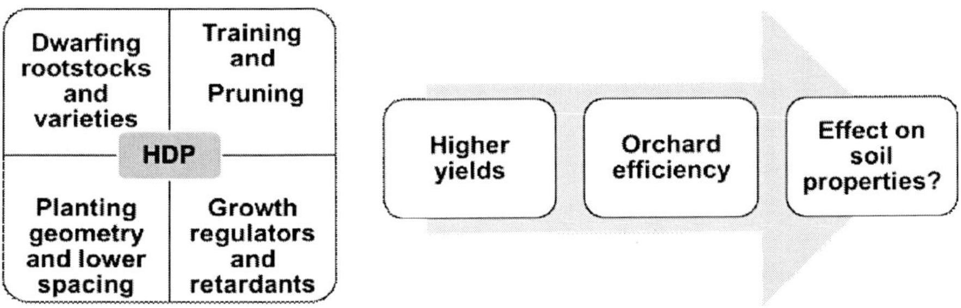

Fig. 1: Drivers of high-density planting (HDP) in mango and its impact

Material and Methods

a) Experimental location and study design

Present investigation was carried at ICAR- Indian Institute of Horticultural Research, Bengaluru, Karnataka. Mango orchard was established in red soils under rainfed conditions located at geo-coordinates 13.13^0 N and 77.49^0 E. Soil properties were assessed in 15 years old established orchard of 'Alphonso' mango grafted on two different rootstocks (r) *viz.* Vellaikolumban and Olour. Three spacing levels (s) i.e. 3 × 3 m, 5 × 5 m and 7.5 × 7.5 m and three paclobutrazol levels (p) i.e. 0, 0.125 and 0.250 g a.i. of paclobutrazol per year age of tree starting from fourth orchard year and stabilized by tenth orchard year. Alphonso variety grafted on unspecified rootstock spaced at 10 × 10 m without any paclobutrazol application served as an external control. So, there were 18 treatment combinations employed in split-split plot design along with control which were replicated twice. Plant population varied from 4 to 25 in uniform block size. Paclobutrazol was drenched in soil at 2 to 3 feet radial distance from tree trunk after harvesting during October, every year.

b) Soil biochemical analysis

Two soil samples were collected randomly from the place of paclobutrazol application at the depth of 0 to 15 cm in each plot. After air drying and sieving, they were stored in plastic bags at room temperature. To estimate soil respiration rate, soil samples were maintained in plastic pots at field capacity condition. Soil respiration was analyzed from 50 grams of soil having average moisture content 23.0 per cent, which was closed in air tight plastic container. Released CO_2 was collected in 0.05M NaOH and titrated against 0.05M HCl with phenolphthalein indicator after addition of 0.5M barium chloride (Isemyer, 1952) and respiration rate was expressed in mg CO_2 kg^{-1} of soil in 24 hours. Dehydrogenase activity was determined by following method of Casida *et al.* (1964) by reduction of 2,3,5- triphenyl tetrazolium chloride (TTC) from one-gram soil incubated for 24 hours at 27^0 C and results were expressed as µg TPF/g of soil/day. Soil glomalin were extracted from one gram of soil as per protocol developed by Wright *et al.*, (1996) and modified by Wright and Upadhyaya, (1998) and glomalin content expressed in mg per gram of soil. Acid and alkaline phosphatase activity assessed with the help of modified buffer solution, p- Nitrophenol phosphate tetrahydrate and calcium chloride solution. Intensity of yellow colour developed was measured at 420 nm after incubation at 37^0 C for one hour. Soil phosphatase activity for both acid and alkaline pH optima was expressed as µg p-nitrophenol per gram of soil per hour.

c) Determination of water infiltration rate

A portable infiltrometer was constructed using two concentric iron rings, each of 60 cm height and having diameters 15 cm and 25 cm, respectively and both rings were driven simultaneously in the soil at least 8 cm deep with help of hammer and wooden poles in midway between two adjacent plants (in-row location of plants). The water was filled in both, inner and outer rings. The fall in water level from inner ring was recorded at 10 minutes interval until it stabilizes for four times consecutively upto 140 minutes. Initial and basic (stabilized) infiltration rates were expressed in mm $hour^{-1}$ while cumulative soil water infiltration was expressed in mm.

Results and discussion

The effect of rootstocks, spacing levels and paclobutrazol on soil biochemical properties and cumulative soil water infiltration is given in Table 1. Soil respiration for 24 hours was 101.20 mg CO_2/ kg of soil in control, which was non-significantly higher than high density mango plots (93.19 mg CO_2/kg of soil). Similarly, there was no impact of rootstocks, spacings and paclobutrazol concentrations on soil respiration rates. Soil dehydrogenase activity, soil glomalin content, acid and alkaline soil phosphatase also not influenced by either rootstocks, planting densities and paclobutrazol levels. There was non-significant impact on these soil enzymatic activities due to paclobutrazol levels.These results are contrary to earlier studies conducted by Silva *et al.*, 2003 and Singh *et al.*, 2005, which has reported significant reduction in soil microbial communityof soil fungi, bacteria and actinomycetes due to residual effect of paclobutrazol. Hence, soil application of paclobutrazol doses calculated based on tree age and tree canopy diameter are proven to be sustainable.

Table 1: Soil biochemical properties and water infiltration rate in high density mango

	Soil Respiration (mg CO_2/ kg of soil /24 hrs)	Soil dehydrogenase activity (µg TPF/g of soil/day)	Soil glomalin (mg/ g of soil)	Acid phosphatase (µg PNP/g of soil/24 hr)	Alkaline Phosphatese (µg PNP/g of soil/24 hr)	Cumulative Soil water infiltration (mm)
Control	101.20[a]	8.33[a]	2.11[a]	796.62[a]	265.62[a]	37.00[b]
Treatments	93.19[a]	6.13[a]	2.08[a]	446.00[a]	269.50[a]	80.25[a]
Rootstocks *(r)*						
Vellaikolumban	91.30[a]	6.39[a]	2.12[a]	455.23[a]	284.86[a]	81.00[a]
Olour	95.08[a]	5.86[a]	2.05[a]	436.70[a]	254.13[a]	80.00[a]
Spacing *(s)*						
3.0 × 3.0 m	91.48[a]	3.86[a]	2.38[a]	366.72[a]	170.95[a]	100.00[a]
5.0 × 5.0 m	89.83[a]	7.68[a]	1.65[a]	485.40[a]	353.69[a]	79.00[b]
7.5 × 7.5 m	98.26[a]	6.83[a]	2.23[a]	485.80[a]	283.85[a]	61.00[c]
Paclobutrazol *(p)*						
0 g	97.16[a]	4.53[a]	2.10[a]	442.45[a]	267.90[a]	74.00[c]
0.125 g	91.30[a]	6.89[a]	2.08[a]	478.47[a]	322.59[a]	81.00[b]
0.250 g	91.11[a]	6.95[a]	2.06[a]	417.00[a]	218.00[a]	86.00[a]
CD value at p<0.05 for soil water infiltration						
r	*s*	*r* × *s*	*p*	*r* × *p*	*s* × *p*	*r* × *s* × *p*
NS	7.01**	9.92**	3.70**	5.23**	6.40**	9.06**

Note: Mean values characterized by same letter within column are not significantly different at (p<0.05)

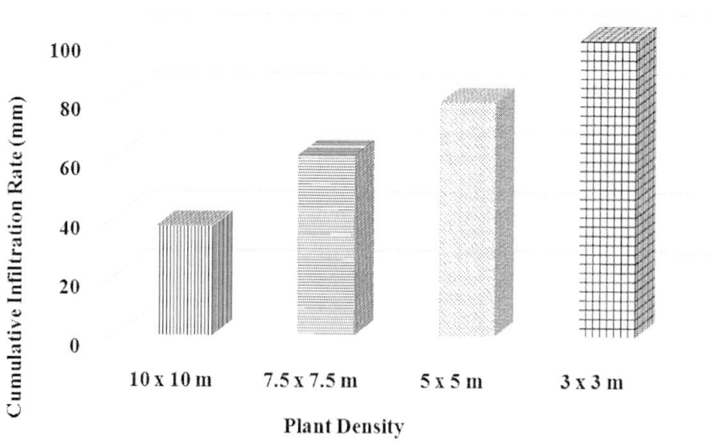

Fig. 2: Cumulative soil water infiltration rate in different plant density

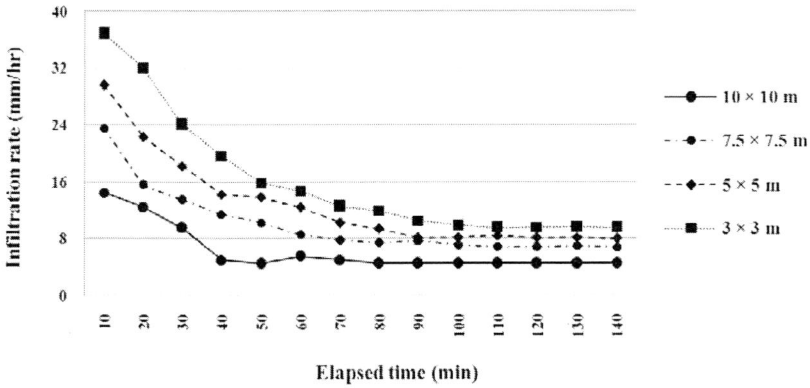

Fig. 3: Initial and basic soil water infiltration rate at 10 minutes interval

Minimum soil disturbances, leaf litter addition and systematic age-based soil application of paclobutrazol in tree basins did not affect soil microbial population and soil biochemical properties eventually. In contrary to results published by Singh *et al.,* 2005 for negative influence of paclobutrazol on soil microbial population, soil application of paclobutrazol doses calculated based on tree age and tree canopy dimeter is proven sustainable.

Basic soil water infiltration was not affected by rootstocks. High density mango treatments plots recorded 53.9% higher cumulative soil water infiltration than control plot of conventional mango planting system (Table 1). Seedling rootstocks or specified rootstocks *viz.* vellaikolumban and olour did not affect initial as well as basic soil water infiltration significantly. Cumulative, initial and stabilized soil water infiltration rates were increased with planting density (Fig 1 and 2). The highest cumulative soil water infiltration of 100 mm was recorded at spacing level of 3.0 × 3.0 m while lowest rate of 37.00 mm was recorded at 10 x 10 m planting distance. Similar trend was observed for initial as well as stabilized soil water infiltration at different densities. The significantly different basic infiltration rates recorded after 140 minutes at spacing distance of 3.0 × 3.0 m, 5.0 × 5.0 m, 7.5 × 7.5 m and 10.0 × 10.0 m were 4.5, 7.0, 8.0 and10.0 mm hour^{-1}, respectively.

Higher soil compaction in widely spaced trees alley might have reduced water infiltration drastically.The improved soil water infiltration caused by high density planting in tree alleys due to higher root density in rhizosphere. This high root density might have increased soil structural stability of this zone which is associated with higher organic matter content and number as well as pore structure in soil profile. Increased number of bio-pores in both rhizosphere as well as subsurface soil of tree alleys has allowed higher soil water intake. Further, the differences between soil bulk density under different planting system may elucidate better the changes in soil porosity. Results of this study emphasizes the importance of crop management practices with reference to intensive horticulture in commercial mango production. Increased infiltration and retention of soil moisture in the rootzone will improve yield through maximized rainfall utilization. Further, sustainability of high-density mango orcharding can be studied with reference to reduction in runoff and soil loss, and potential of groundwater recharge under different planting density and geometry.

Acknowledgement

Authors are thankful to the Director, ICAR-IISWC, Dehradun and the Director, ICAR-IIHR, Bengaluru to complete this study as a part of Professional Attachment Training at ICAR-IIHR, Bengaluru during May to August, 2014.

References

Casida Jr, L.E., Klein, D.A. and Santoro, T. 1964. Soil dehydrogenase activity. *Soil Science*, 98(6): 371-376.

Isemeyer, H. 1952. Eine einfachemethodezurbestimmung der bodenatmung und der karbonateimbodem. In Kassem Alef and Paolo Nannipieri. Eds. Methods in Applied Soil Microbiology and Biochemistry. New York, Academic Press: 215

Kotur, S.C., 2006. Effect of paclobutrazol on root activity of mango (*Mangiferaindica*)." *Indian Journal of Agricultural Science* 76(3): 143-144.

NHB, 2016. Horticultural Statistics at Glance, 2015.Horticulture Statistics Division Department of Agriculture, Cooperation & Farmers Welfare Ministry of Agriculture & Farmers Welfare, Government of India

Rademacher, W., 1995. Growth retardants: biochemical features and applications in horticulture. *ActaHort.* 394: 57-74.

Silva, C.M.M.S., Vieira, R.F. and Nicolella, G. 2003. Paclobutrazol effects on soil micro-organisms. *Applied Soil Ecology*, 22(1): 79-86.

Singh V.K., Garg N and Bhirigwanshi S.R. 2005. Effect of paclobutrazol doses on nutritional and microbiological properties of mango (*Mangiferaindica*) orchard soils. Indian Journal of Agricultural Sciences 75: 738–9.

Wright S.F., Franke-Snyder M., Morton J.B. and Upadhyaya A. 1996. Time-course study and partial characterization of a protein on hyphae of arbuscular mycorrhizal fungi during active colonization of roots. Plant Soil, 181: 193–203.

Wright, S.F. and Upadhyaya, A. 1998. A survey of soils for aggregate stability and glomalin, a glycoprotein produced by hyphae of arbuscular mycorrhizal fungi. *Plant and Soil*, 198(1): 97-107.

❑❑❑

Chapter 13

Effect of *in-situ* Moisture Conservation on Establishment of Tree Borne Oilseeds (TBOs) Plantation in Red Soils of Bundelkhand Region

Manish Kumar[1,4*], Monalisha Pramanik[1], Rajeev Ranjan[1], M.N. Ramesha[3], Prabhat Kumar[2] and Raj Kumar[4]

[1] ICAR-Indian Institute of Soil and Water Conservation, RC, Datia - 475661; [2] ICAR-National Research Centre on Litchi, Muzaffarpur-842002, Bihar, [3] ICAR-Indian Institute of Soil and Water Conservation, RC, Bellary-583101, [4] ICAR-Cental Soil Salinity Research Institute, Karnal-132001, Haryana, *E-mail-manish.rrm@gmail.com

ABSTRACT

Tree borne oilseed crops (TBOs) are known to be multipurpose trees in agroforestry systems to diversify the income of farmers and provides various ecological services. Production of TBOs on wastelands meets the twin objectives of rehabilitation of a vital land resource and obtaining an energy substitute. Extensive wastelands of Bundelkhand in the form of ravine tracts, undulating terrain with scarce vegetation cover, rocky outcrops and scrublands are the results of continued soil degradation and biotic interference is suitable for large scale plantation of indigenous TBOs. Plantation of three TBOs, namely Neem (*Azadirachta indica*), Karanj (*Pongamia pinnata*) and Mahua (*Madhuca indica*) with moisture conservation measures *viz.* trenches (double and single), V-shape catchments and simple pits were established in 2011 at research farm of ICAR-IISWC, Datia under red soil of Bundelkhand region for its growth evaluation along with suitable soil-moisture conservation measures. Results revealed that *P. pinnata* showed better growth performance with respect to all biometric parameter *viz.* height, collar diameter, number of branches, diameter at breast height (DBH) and crown spread in comparison to other species. Significantly, higher growth of *P. pinnata* in terms of plant height, collar diameter, number of branches and canopy spread (372.7 cm, 9.25cm, 17.2 and 327.9 cm, respectively) were recorded, followed by *A. indica* and *M. indica*. Soil moisture data suggested the superiority of "double trench treatment" over other moisture conservation practices, in conserving the *in-situ* moisture content. Significantly, better *in situ* moisture content (10.05% and 12.44%) with double trench followed by single trench, V-shape catchment while minimum in ordinary pits. Different soil-moisture conservation measures improve moisture availability during critical stages of growth and development by directing and concentrating rainwater through runoff to the plants. TBOs establishment on degraded or wasteland is not an easy task, while trench (double and single) techniques of soil moisture conservation is effective for establishment and growth of TBOs on degraded land of Bundelkhand region.

Keywords: Bundelkhand, Growth performance, Moisture conservation techniques, Tree Borne Oilseeds (TBOs)

Introduction

Energy and food security along with environment degradation is main cause of concern for developing counties. All governments across the globe are working towards green economy and tried to reclaim the barren and unproductive land through green cover. Green cover on barren and wasteland will reverse the process of degradation mainly caused by water and soil erosion, and fertility of these unproductive lands will improve through recycling of nutrients from deeper layers, addition of leaf litter, nitrogen fixation trees like *Pongamia pinnata* (Wani and Sreedevi, 2007). Perennials, such as *P. pinnata* have offered an excellent opportunity to remedy the problems of environmental protection and oil crisis. Strategies of rehabilitating degraded lands and improving livelihoods through biodiesel plantations (Wani *et al.*, 2006.) is good initiative steps in developing countries. The use of green fuel will stop degradation of land, environment and also improve the rural livelihoods. This is also likely to reduce the dependence on fossil fuel and ensure energy security in future (Wani and Sreedevi, 2007). Greening of unproductive lands and regeneration of degraded forestlands could be achieved through establishment of indigenous tree bearing oilseeds (TBOs). Adoption of TBOs multi-purpose tree species in agroforestry systems supports to diversify the income of farmers and provides various ecological services. Production of TBOs on wastelands meets the twin objectives of rehabilitation of a vital land resource and obtaining an energy substitute. This is a highly appealing combination because it includes a number of ecological and economic benefits.

Bundelkhand region (7.04 M ha) of central India has undulating terrain with scarce vegetation cover, harsh climate and unfavourable edaphic conditions. The productivity of red soils (alfisols), which comprises more than 50 per cent area in the region, is very low due to low water holding capacity, low fertility and limiting soil depth and even crop failure due to frequent drought and long dry spells during rainy season. Extensive wastelands in the form of ravine tracts, rocky outcrops and scrublands are the results of continued soil degradation and biotic interference. Establishment of vegetation and growth is slow due to the predominance of severe moisture stress conditions caused by low and sporadic rainfall. The semi-arid lands are characterized by low and unreliable rainfall, with lack of soil moisture which limits to vegetative growth. Therefore, moisture conservation measures are essential in semi-arid tract for supporting establishment of trees. Approximately, 50-63% of the forest in this region is degraded and should be rehabilitated by adopting *in-situ* moisture conservation measures and planting of indigenous and locally available land races of fruit trees, fodder trees, shrubs, grasses (NRAA, 2008). The water availability for tree survival and growth can be improved through various soil and water management practices. V-shaped micro-catchment could be a suitable *in-situ* moisture conservation practice for resource conservation and enhancing growth and yield of aonla (*Emblica officinalis*)under agroforestry system on sloping lands in red soils of Bundelkhand region (Narayan *et al.*, 2017; Kumar *et al.*, 2018). The percentage increase in fruit yield of aonla over farmer's practice was observed by 8%, 13% and 40% with 75% pits, crescent shaped and V-shaped micro-catchment, respectively (Kumar *et al.*, 2018). Rainwater harvesting and moisture conservation treatments influenced the soil moisture storage, growth and biomass production of *A. indica* in the Indian desert (Gupta, 1994). The soil and moisture conservation techniques not only conserve the soil moisture but also check the soil erosion, which arrests the further land degradation.

Azadirachta indica (Neem) is an indigenous tree having the potential of withstanding dry climatic conditions and can grow in tropical and subtropical regions with semi-arid to humid climates. *Pongamia pinnata* (Karanj) is a drought tolerant, semi-deciduous tree and a potential source of biodiesel (Kureel *et al.*, 2008; Scott *et al.*,

2008). The alternative fuel search has led to *Madhuca indica* (Mahua) oil as diesel fuel in India (Kapilan and Reddy, 2008). Thus, local TBOs *viz., M. indica, A. indica* and *P. pinnata* were chosen for the study. The assessment of moisture conservation methods may be imperative to address the problems of large scale afforestation programme on wasteland. Thus, Indigenous tree species needs to be evaluated in terms of early plantation establishment, growth performance and economic viability in Bundelkhand region.

Methodology

A field experiment was conducted at ICAR-Indian Institute of Soil and Water Conservation, Research Centre, Datia, Madhya Pradesh (78° 26′ 04 E and 25° 42′ 14 N). Soil of experimental site comes under red soils (alfisol), which are highly heterogeneous in texture, depth and fertility status, holds less water content and often fail to support crop on residual soil moisture. Bench mark of soil profile study and sampling was done. Climate of the region is hot semi-arid. Average annual rainfall is approx. 800 mm with high degree of spatial and temporal variations. The region has short rainy season, erratic rainfall, winter temperature plunges down to -1°C and summer boils upto 47°C. Water stress conditions are common even in rainy season with high annual evaporation losses ranges between 1400 and 1700 mm.

The experiment was laid in 2011-12 with spilt plot design, accommodating three TBOs, namely Neem (*Azadirachta indica*), Karanj (*Pongamia pinnata*) and Mahua (*Madhuca indica*) in main plots and four moisture conservation measures *viz.*, trenches (double and single), V-shape catchments and ordinary pits in subplots and replicated thrice. Pits having dimension of 0.6m depth, 0.45m width and 0.45m length were dug. Planting of uniformly graded saplings was done after onset of monsoon in the month of June 2011. Proper basins around the saplings were prepared to collect the runoff from moisture conservation measures to the saplings. To enrich soil organic matter, 5 kg farmyard (FYM) manure was applied to each pit before planting. No manuring was done afterwards to mimic afforestation works pattern usually carried out in India.

Growth parameters and soil moisture

Tree biometric parameters, such as survival per cent, plant height, collar girth, number of branches and crown spread were recorded in the 4-5th year of plantation during 2015-16. Plant height was recorded by using measuring pole marked in metre and collar diameter/diameter at breast height (DBH) with the help of digital caliper at a height of 10-15 cm above the ground level. Canopy spread was measured with the help of measuring tape in east-west and its perpendicular direction. *In-situ* soil moisture content was recorded with moisture meter (Field Scout TDR-200, Spectrum-Tech. Inc.) at the depth of 20 cm within radius of 15-25 cm distance from the saplings.

Statistical analysis

The data were analyzed statistically as per the standard procedure by using online design resource server, IASRI, New Delhi (www.iasri.res.in/design) and means compared by the Duncan's multiple range tests (DMRT) at the d"0.05% level of probability for interpretation of results and drawing conclusions.

Results and Discussion

Data on different moisture conservation techniques for collecting the runoff to enhance *in situ* soil moisture content were recorded during summer (May-June) and winter (Oct-Nov). *In situ* soil moisture content significantly differs among the various moisture conservation techniques as well as under the different tree canopy before monsoon (Table 1). However, soil moisture content did not exhibit any variation with different moisture conservation techniques during rainy season. Double trench, single trench, and V-shape catchment moisture conservation treatments were superior and retained significantly higher *in situ* soil moisture content over ordinary pit planting. However, supplemental life saving irrigation needs to be given for the saplings during summer in the region to cope up hot and dried condition. Double trench moisture conservation techniques showed significantly better *in situ* moisture content (10.05 and 12.44, during summer and winter, respectively) followed by single trench, V-shape catchment while minimum in ordinary pits. The findings of present study are in accordance with earlier findings of Ramesha *et al.* (2014). Soil moisture content highly variable with time, site, canopy and trenching, as well as general precipitation patterns (Powell and Bork,

2007). Rainwater harvesting techniques are essential for crop production, especially in arid and semi-arid regions (Komariah and Senge, 2013). These results confirm the findings of several workers who also agree with influence of rainwater conservation techniques adopted by Li *et al.* (2005; 2006) and Gupta (1994) for raising *Tamarix ramosissima* on loess in China and *A. indica* in the Indian desert, respectively. Rain-water harvesting can improve moisture availability for agriculture by directing and concentrating rainwater through runoff to the plants and other beneficial uses. The socio-economic & environmental benefits of this practice are far more important than increasing agricultural water productivity (Oweis and Hachum, 2006).

Table 1: *In-situ* soil moisture as influenced by various soil moisture conservation techniques

Treatment	In-situ soil moisture content (%) in 2015	
	Before monsoon (May-June)	After monsoon (Oct-Nov)
A. Tree species		
M. indica	10.08[a]	9.96
A. indica	8.48[b]	9.61
P. pinnata	8.04[b]	10.74
CV (%)	15.03	9.69
B. Moisture conservation		
Single trench	9.49[a]	10.96[b]
Double trench	10.05[a]	12.44[a]
V-shape catchment	8.89[a]	9.55[c]
Ordinary pit	7.04[b]	7.47[d]
CV (%)	18.41	10.05
Interaction (AxB)	Non-Significant	Non-Significant

Letter ([a,b]) indicates significant differences at 5% level of significance ($p < 0.05$)

The growth parameters in terms of plant height, collar diameter., canopy spread and number of branches of TBOs, such as *M. indica, A. indica* and *P. pinnata* influenced by adopting different soil moisture conservation treatment over ordinary pits planting (Table 2). The plant growth in terms of plant height, collar diameter, crown diameter and DBH of *P. pinnata* has been found better than other two species (Fig. 1). The trenches (double and single) were more effective than ordinary pits. Plant treated with double trench techniques recorded highest growth parameters, plant height, collar diameter, number of branches and canopy spread (292.4 cm, 6.29cm, 11.2, 174.7 cm, respectively) followed by single trench, V-shaped catchments, while minimum in Ordinary pits. Among tree species, *P. pinnata* performed better than other two species in all the growth parameter. *P. pinnata* showed the significant growth in terms of plant height, collar diameter, number of branches and canopy spread (372.8 cm, 9.25cm, 17.2 and 327.9 cm, respectively) over *A. indica* and *M. indica*. Various *in-situ* soil moisture conservation practices helped in conserving higher rainwater, better moisture availability that in turn resulted in higher growth of *P. pinnata* in comparison to ordinary pit planting. Similarly, *In-situ* moisture conservation was found to be effective in improving the tree survival and growth, canopy spread in bael by Shukla *et al.*, (2014), aonla (Narayan *et al.*, 2017; Kumar *et al.*, 2018) and *A. indica* in the Indian desert (Gupta, 1994).

Fig. 1: Performance of *P. pinnata* in terms of different plant growth parameter was found better than other two TBOs species *M. indica* and *A. indica*

Table 2: Effect of different soil moisture conservation treatment on plant growth parameter

Treatment	Plant growth parameter			
	Plant height (cm)	Collar diameter/DBH* (cm)	No of branches	Canopy spread (cm)
A. Tree species				
M. indica	209.3[c]	4.85[b]	7.2[b]	67.0[b]
A. indica	279.1[b]	4.11[b]	9.3[b]	96.4[b]
P. pinnata	372.8[a]	9.25[a]	17.2[a]	327.9[a]
CV (%)	16.6	24.88	49.76	31.5
B. Moisture conservation				
Single trench	295.3	6.19	11.9	171.7
Double trench	292.4	6.29	11.2	174.7
V-shape catchment	282.1	5.91	10.7	158.4
Ordinary pit	278.4	5.89	10.2	160.4
CV (%)	10.08	11.19	16.74	13.34
Interaction (AxB)	NS	NS	NS	NS

NS=Non-Significant, Letter ([a,b]) indicates significant differences at 5% level of significance ($p < 0.05$)*DBH in case of *P. pinnata*

The Current annual increment (CAI) of different TBOs, such as *M. indica, A. indica* and *P. pinnata* were influenced by adopting different soil moisture conservation treatment over ordinary pits planting (Fig.1). CAI in collar diameter was highest with double trench in TBOs, followed by single trench, V-shape catchment while lowest with ordinary pit (control). The highest CAI in collar diameter was recorded with *P. pinnata* in comparison to other *A. indica* and *M. indica*. Rain-water harvesting system like full moon, half moon and trench in combination with mulch improves growth and fruit yield of apple under rainfed conditions by enhancing the availability of soil moisture during critical stages of fruit growth and development (Kumar *et al.*, 2014).

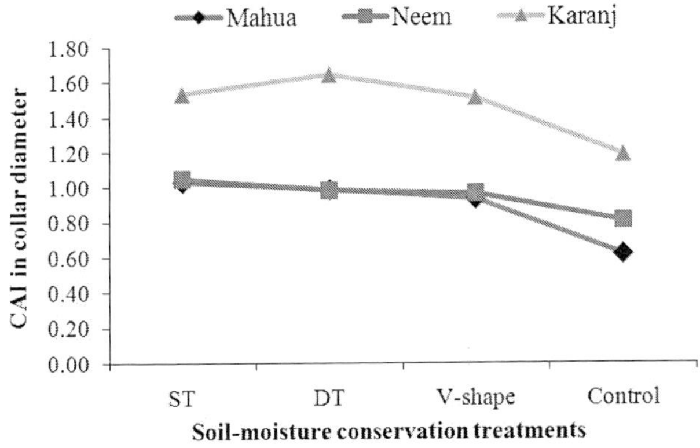

Fig. 2: Current annual increment (CAI) of collar diameter as influenced by various soil moisture conservation treatments

Conclusion

Establishment of TBOs plants on degraded land of arid and semi-arid region of Bundelkhand are relatively difficult, where availability of soil moisture to the plants remains always a constraint. In these regions, application of soil conservation techniques like construction of trenches (single and/or double trench) are found to be an effective way for harvesting the runoff to enhance *in-situ* soil moisture content and remain available to the plants for longer duration. This played an important role in proper growth and establishment of TBOs on degraded land of Bundelkhand region. The present observation suggested that, out of three TBOs (*A. indica, M. indica* and *P. pinnata*), *P. pinnata* performed better even in long dry spells with different soil moisture conservation. The survival and growth performance of *P. pinnata* were recorded better than that of *A. indica* and *M. indica*. Thus, construction on trenches is a cost effective and economically viable technology for establishment of TBOs plantation on degraded land of Bundelkhand region. However, more experimentation is required to quantify the growth performance on different soil types for wide acceptance.

Acknowledgement

The authors are sincerely grateful to Director, ICAR-Indian Institute of Soil and Water Conservation (IISWC), Dehradun and and Head, ICAR-IISWC, Research Centre (RC), Datia for providing the necessary facilities to conduct the study.

References

Gupta, G.N. 1994. Influence of rain water harvesting and conservation practices on growth and biomass production of *Azadirachta indica* in the Indian desert. *Forest, Eco Manage.* 70(1-3): 329-339.

Indian Agricultural Statistics Research Institute (ICAR), New Delhi 110012, India. www.iasri.res.in/design (accessed lastly on 01.11.2018).

Kapilan, N. and Reddy, R.P. 2008. Evaluation of Methyl Esters of Mahua Oil (*Madhuca indica*) as Diesel Fuel. *J Am Oil Chem. Soc.* 85: 185-188.

Komariah, and Senge, M. 2013. The development of water harvesting Research for agriculture. *Reviews in Agricultural Science.* 1: 31-42. doi: 10.7831/ras.1.31

Kumar, D., Ahmed, N. and Hussan, A. 2014. Impact of rainwater harvesting and mulching on growth and yield of apple (Malus domestica) under rainfed condition. *Indian Journal of Soil Conservation,* 42(1): 74-79.

Kumar. M., Narayan, D., Yadav, R.S., Kumar, P., Ranjan, R., Pramanik, M. and Raj Kumar, 2018. Performance of Aonla with *in-situ* Moisture Conservation Techniques. *Indian Journal of Ecology,* 45(3): 654-658.

Kureel, R.S., Singh, C.B., Gupta, A.K. and Ashutosh P. 2008. Karanja-A potential source of Bio-diesel. NOVOD Board, Gurgaon, India. p.22.

Li, X.Y., Liu, L.Y., Gao, S.Y., Shia, P.J., Zou, X.Y. and. Zhang, C.L. 2005. Microcatchment water harvesting for growing *Tamarix ramosissima* in the semi-arid loess region of China. *Forest Ecology and Management,* 214: 111–117.

Li, X.Y., Shia, P.J., Sun, Y. L., Tang, J. and Yang, Z.P. 2006. Influence of various *in-situ* rainwater harvesting methods on soil moisture and growth of *Tamarix ramosissima* in the semi-arid loess region of China. *Forest Ecology and Management,* 233(1): 143-148.

Narayan D., P. Kumar, Kumar M. and Yadav R.S. 2017. Rainwater harvesting practices and resource conservation in agroforestry system in red soils of Bundelkhand. *Indian Journal of Agroforestry,* 19(2): 34-37.

NRAA, 2008. Report on Drought Mitigation Strategy for Bundelkhand Region of Uttar Pradesh and Madhya Pradesh. Ministry of Agriculture, Government of India, New Delhi, pp. 144.

Oweis, T. and Hachum, A. 2006. Water harvesting and supplemental irrigation for improved water productivity of dry farming systems in West Asia and North Africa. *Agricultural Water Management*, 80: 57-73.

Powell, G.W. and Bork, E.W. 2007. Effects of aspen canopy removal and root trenching on understroy micro-environment and soil moisture. *Agro Forest Syst.* 70: 113-124. doi 10.1007/s10457-007-9051-z.

Ramesha, M.N., Adhikary, P.P., Dwivedi, V.K., Kumar, P. and Tiwari, S.P. 2014. Challenges for establishment of oilseed bearing trees in semi-arid tropics of Bundelkhand, India. *Eco. Env. & Cons.* 20 (Suppl.): 213-223.

Scott, P.T., Pregelj, L., Chen, N., Hadler, J.S., Djordjevic, M.A. and Gresshoff, P.M. 2008. *Pongamia pinnata*: An untapped resource for the biofuels industry of the future. *Bioenerg. Res.* 1: 2-11.

Shukla, A.K., Kumar, S., Ram, S.N., Singh, H.V., Watpade, S.G and Pramanick, K.K 2014. Bael (*Aegle marmelos*) based horti-pastoral system with moisture conservation in semi-arid condition. *Journal of Tree Sciences,* 33(1): 7-11.

Wani, S.P. and Sreedevi, T.K. 2007. Strategy for Rehabilitation of Degraded Lands and Improved Livelihoods Through Biodiesel Plantations. *In*: 4th International Conference on Biofuels, February 1-2, 2007. Winrock International, India.

Wani, S.P., Osman, M., D'Silva, E. and Sridevi, T.K. 2006. Improved Livelihoods and Environmental Protection through Biodiesel Plantations in Asia. *Asian Biotechnology and Development Review,* 8(2): 11-29.

❑❑❑

Chapter 14

Integrated Organic Farming System: Technology Generation and Dissemination for Sustainable Hill Agriculture

***Anup Das, Subhash Babu, Jayanta Layek, Ramkrishna G.I., Thoi Thoi Devi, Dauni Suiting, Utpal Dey, GS Yadav and S.V. Ngachan**

ICAR Research Complex for NEH Region, Umiam, Meghalaya *Email: anup_icar@yahoo.com*

Introduction

The north-eastern region (NER) of India has vast opportunity to emerge as major suppliers of organic products. India had brought more than 5.21 million hectare area under organic certification process by 2013-14 (APEDA). The world market for organic products was estimated more than US $ 25 billion in 2003 and has risen by about 20% per year during 2001 to 2005 (Willer and Yussefi, 2005). Economic and environmental concerns have been the drivers for increasing demand for organically produced food. Due to growing purchasing power and health cautiousness, organic food products are being preferred by the consumers and demand for organic vegetables is steadily increasing not only in international markets but also in domestic markets. Organic farming (OF) is considered as one of the best options for protecting/sustaining soil health and produce healthy foods (Das *et al.*, 2017). The soil fertility is maintained by returning all the residues to soil through composts and residue recycling, thereby, minimizing the gap between nutrient addition and removal from the soil. When farming components are identified considering the agro-climatic conditions, farmers need, market demand; organic farming can provide much higher income than conventional practice due to higher price. To make the organic agriculture sustainable, it is necessary to utilize all the resources available on- and off the farm effectively by integrating complementary and supplementary enterprises. With this background, Integrated Organic Farming System (IOFS) has been envisaged for soil and water conservation for sustainable hill agriculture.

Land-Use Scenario of the Region

The most of the states in the NER have large proportion of land under forests. In general, over 54% of the geographical area of the region is recorded as notified forests - reserve, protected and un-classed forest under the control of forest department. The net and gross cultivated area in NER is about 4 m ha and 5.7 m ha, respectively. The share of total cultivated area under food grain production as against total cropped area is highest in Manipur (78.7%), followed by Nagaland and Arunachal Pradesh. Except Sikkim and Tripura, all other NER states are having higher share of area under food grain production compared to national average (65%). This indicates that there is scope for diversification of food crop based cropping systems.

In NER, around 56% of the area is under low altitude (valley or lowland), 33% mid-altitude (flat-upland) and the rest under high altitude (upland terrace). Traditionally, farmers both at upland terrace and valley land follow the mono-cropping practice in rainfed agriculture, where rice is the major crop occupying more than 80% of the cultivated area followed by maize. The NER of India has 8 distinct agro-climatic zones starting from subtropical to temperate to Alpine zone. Each state has been again divided into agro-ecozones for agricultural planning and development. Rice is the main crop of Arunachal Pradesh and tropical and temperate fruits are also grown. In Assam, double cropping of rice is practiced in plains. Fish farming is also common. Tea husbandry is common in hills of Assam and Tripura. In Manipur, rice, oilseeds, vegetables are major crops. Rice, maize, ginger, turmeric, citrus, etc. are important crops in Meghalaya. Rice is grown on terraces of Mizoram with horticultural crops, such as orange on slopping areas. Rice is cultivated in valleys and horticultural/plantation crops on hills of Nagaland. Sikkim is only state where agriculture is well established on bench terraces and no *jhum* cultivation is prevalent. Maize and cardamom are the important crops of Sikkim. Sikkim is the only state in India which is an organic state. In Tripura, there are three agro-eco-regions where double cropping of rice is prevalent in plains. Pineapple, arecanut, and vegetables are the other important crops.

Opportunity of OF in NER

The North-Eastern Hill region of India has numerous advantages to go for organic food production due to minimum use of fertilizer, availability of plant- and livestock-excreta-based organic manure and favourable climatic conditions for growing a wide range of crop species.The use of inorganic fertilizers and chemicals is meager in the region. The farmers of the region, in general and hill farmers in particular are having apathy towards use of agro-chemicals. Further, the fruits of green revolution could not benefit the farmers of the hills, as the system of production in the hills remained low input-low risk-low yield technology and the average yield of most of the crop remained far behind. It is assumed that the difference in production gap due to adoption of organic agriculture is expected to be negligible in NEH region; rather there is scope for enhancing productivity with good organic management and the organic premiums would boost earning of the hill farmers. Thirdly, it is an added advantage that all the households are maintaining livestock (pig, poultry, cattle, goats, etc.) and the region is having about 26 million livestock producing sufficient quantity of on-farm manures, which could be efficiently used for organic agriculture. Moreover, the north-eastern states being the one of the mega biodiversity receiving very high rainfall (2000 mm to 11000 mm per annum) leads to profuse production of biomass including weeds, shrubs and herbs. Some of these species could be efficiently used in organic production. The region as a whole having a potential of 46 million tonnes of manure including crop residues and livestock excreta's, (Bujarbaruah, 2004).Considering the cultivated area of the region at little above 4 million hectare, this quantity of manure and biomass is a good amount for going organic farming (Das *et al.*, 2014).

Crop residues account for 1.66 times of rice grain, 2.32 times of maize grain, 1.66 times of pulses and 1.33 times of oilseeds. The total potential for nutrient supply through crop residues in the region have been estimated as 9.86, 2.12 35.5 tons of NPK, thus, creating on an average 2.46 kg N, 0.53 kg P_2O_5 and 8.87 K_2O/ha. Thus, a good quantity of potash can be added through crop residues (Sharma *et al.*, 2006).

Some other strengths of north-eastern region for organic farming are :

- The region is home to some high value niche crops like Assam lemon, Joha rice, medicinal rice, sticky rice (Manipur), aromatic rice of Tripura (*harinarayana* & *kalikhasa*), Assam tea, large cardamom, ginger, pine apple and passion fruits which has high market demands.
- Less pressure on food demand due to thin population density as compared to vast natural resources available in the region.
- Dependence of mid and high altitude farmers on within farm renewable resources.
- Time tested indigenous farming systems and use of indigenous technical knowledge in agriculture are mostly prevailing in the region.
- Agri-Export Zones (AEZ) has already been established for organic cultivation of pineapple in Tripura and for ginger in Sikkim.
- Do NER Ministry, Govt. of India has a special 100 crore project for promotion of organic farming in the NER during 12 five year plan.

Organic Farming in Mid-altitude of North-East India: some experiences (2004-16)

To assess impact of organic farming on productivity and environmental quality research on organic farming was initiated under Network Project on Organic Farming (NPOF) since 2004. Some of the valuable research data generated from the study are indicated below:

- Supply of nutrients on the basis of N equivalent through integrated *i.e.*, 50% recommended doses of N through FYM + 50% recommended doses of N through VC was found best for improving productivity and soil health in vegetable (french bean, potato and tomato) based cropping systems. System productivity in terms of maize equivalent yield (MEY) was found maximum in maize + soyabean –french bean (35.75 t/ha) cropping system. Soil properties, such as bulk density (1.09 g/cc), soil organic carbon (2.33%) and soil microbial biomass carbon (222.50 μg/g dry soil) showed maximum improvement under integrated nutrient management after eight cropping cycle.

- The results obtained after eight cropping cycle, revealed that rice-tomato cropping system was found most profitable, followed by rice - carrot compared to all other systems and integrated sources of nutrient supply was found more profitable, followed by organic and inorganic sources of nutrients on raised beds.
- Produce quality of tomato, such as TSS (4.70%), ascorbic acid (28.06 mg/100 g), reducing sugar (2.53%), total sugar (4.81%) and lycopene (16.49 mg/100 g) was found better under organic management practice, followed by integrated sources of nutrient supply. Quality of carrot, such as TSS (8.28%), ascorbic acid (39.94 mg/100 g), acidity (0.21%), beta-carotene (7.97 mg/100g) and total carotenoids (68.93 mg/100g) were found better under organic management practice, followed by integrated sources of nutrient supply.
- After eight cropping cycles, soil organic carbon increased by 25.5% and 20.7% compared to the initial value of 2.5% in integrated and organic treatment, respectively. The SMBC was found maximum (222.50 μg/g dry soil) in integrated treatment.
- Field experiments on plant protection revealed maximum yield of maize and soyabean with the application of derisom (3 ml/lit) alone or in combination with panchagavya (3%) and cow urine (10%), whereas, maximum fruit yield of tomato was recorded with panchagavya (3%) + lantana leaf extract (10%) + vermiwash (10%). Sole application of panchagavya (3%) or in combination with lantana leaf extract (10%) and vermiwash (10%) was found effective in minimizing the per cent disease index (PDI) and per cent insect damage in maize, soybean and tomato. Application of neem oil (3 ml/lit) and derisom (3 ml/lit) was found effective in controlling stem borer, Monolepta and Epilenchna in maize and leaf folder, flea beetle and mylloceros in soyabean.
- So far package of practices for 32 crops have been developed and the technology generated through experimentation have been disseminated to farmers field in an integrated farming system approach involving the community.

Development of IOFS model at ICAR RC NEH, Umiam

An IOFS model (0.43 ha) involving crops *viz.*, cereals (rice, maize), pulses (pea, lentil), oilseeds (soyabean), fruits (lemon and papaya), vegetables (French bean, tomato, potato, carrot, cabbage, cauliflower, broccoli, chilly, brinjal, coriander, spinach), dairy (1 cow and one calf), fishery (in farm pond), vermicomposting unit, fodder block, *etc.* have been developed under valley condition. The average results indicated that the rice equivalent yield (REY) from the IOFS model (0.43 ha) was 8.54 t (19.8 t/ha) as against 1.72 t (4 t/ha) and 2.90 t (6.72 t/ha) from the farmers practice - I and II, respectively. The enhancement in net income due to IOFS was 355 and 191% relative to farmers practice-I and II, respectively. The model could meet 92, 82 and 96% (N, P_2O_5 and K_2O) of its nutrient demand within the system. Remaining nutrient demands were met through the external resources like neem cake, rock-phosphate, bio-fertilizer, etc. The results revealed that REY and net returns of the system enhanced by more than 4 times compared to that under farmers practice of mono-cropping of rice. The net income from the IFS model was Rs. 77,442 or Rs. 5953 month^{-1} or 198 day^{-1} which is modest amount for living by a four member family (2 adults and 2 children).

IOFS in Cluster Approach- A Case Study

The North-Eastern Hill region of India has numerous advantages to go for organic food production due to minimum use of fertilizer, availability of plant- and livestock-excreta-based organic manure and favourable climatic conditions for growing a wide range of crop species. A village in Ri-Bhoi district of Meghalaya, namely Mynsain have been adopted for disseminating organic production technology developed in the ICAR Research Complex for NEH Region, Umiam through a model village concept under Network Project on Organic Farming-Tribal Sub Plan (NPOF-TSP) with financial assistance from ICAR (Institute of Farming System Research, Modipuram).

The village is having 132 households with an approximate area of 60 ha. A sensitization meeting with the villagers including village head (Headman), member of the SHGs, Department of agriculture (Gram Sabah) as well as Participatory Rural Appraisal (PRA) was conducted to collect basic information about the village with regards to resources available, type of crops grown, soil quality, livestock, land-use, productivity, problems, etc. for work out the plan of activities. PRA revealed that the villagers are mostly followed mono-cropping of rice or maize, cultivate low yielding local varieties of vegetables and thus, generated very poor farm income. Free grazing during winter season limits the scope for double cropping. As per the interaction with the farmers and elderly people of the village, it is learnt that the village is totally organic and so far no inorganic input has been applied.

Five new ponds (about 400 m^2 area and 1.5 m depth each pond) were constructed and seven existed ponds were renovated in farmer's field of Mynsain village for multiple uses like irrigation and composite fish culture. A total of 28 small rainwater harvesting structure - *jalkund* having 30,000 liters capacity, each were developed for growing vegetables, such as French bean, cabbage, broccoli, tomato, lettuce, cucurbits and for rearing of animals, such as pig and poultry specially during dry season. A community vermicomposting unit (size 6m x 8m x 2.6m) consisting of eight tanks (size 2m x 1.5m x 0.75m) has been constructed in the village with an objective to produce vermin compost by recycling farm biomass. Raised and sunken beds with 1:1 dimension were developed in 10509 m^2 area after rice harvest in lowland for cultivation of vegetables like tomato (var. Avinash, Rocky), French bean (var. Naga local), potato (var. Kufri Megha), carrot (var. New Kuroda), lettuce, etc. Six hundred numbers of improved varieties of Guava seedlings were planted in farmer's field covering an area of about 4500 m^2. Multi-purpose trees along with fodder were also grown for rehabilitation of degraded land and supply of fodder to cattle in lean period. Under the program, seeds of improved verities of crops and vegetables, planting materials, lime, rock phosphate, neem cake and other organic inputs were provided to the adopted farmers. For effective soil fertility management application of weed, decomposed farmyard manure, pig manure, green leaf manure, composts, liming, rock phosphate, etc. were promoted. For pest and disease management, use of neem oil, trichoderma, derisom, indigenous technical knowledge was emphasized. Crop rotation with legumes, such as groundnut, soyabean, French bean, etc. were introduced. To promote small scale mechanization, implements and tools like paddy thresher, cono-weeder, sprayer, rose cans, maize sheller and electric pump has been provided to the village. The farmers were given training in various aspects of organic farming along with conservation of natural resources and residue recycling. Successful cultivation of pea, rapeseed and lentil were followed in rice fallow under no-till. Farmers were provided with improved breeds of pig (75% Hampshire and 25% mixed local) and poultry (Vanaraja) for higher productivity, nutritional security and income. Mostly the farmers were doing mixed cropping of ginger-colocasia–chili for higher income. Five farmers in Mynsain village have already started practicing organic farming in integrated farming system (IFS) mode. They integrated crops (rice, maize), vegetables (tomato, French bean, potato, lettuce, and carrot), livestock (dairy/ piggery), water harvesting (*jalkund*), etc. in IFS mode. Mrs Hynniew Rynghang, an adopted farmer of the village narrated that 'The integrated organic farming system started giving very good returns with very less quantity of external inputs'.

As the villagers were not using any synthetic fertilizers or pesticides earlier, the chance of reduction in yield due to adoption of organic farming does not arise. Rather due to adoption of improved organic production technology, the yield of rice, maize, French bean, ginger, tomato, carrot and chilly had been enhanced by about 15, 22, 40, 33, 45, 37 and 27%, respectively over conventional practice. Villagers are currently selling their produce in local market and along the highway side as uncertified organic produce with 10-15% higher market price as compared to conventional produce. Interaction with the farmers revealed that on an average, the incomes of the farmers are increased by about Rs. 10,000/annum due to adoption of integrated organic farming practices. It is expected that, with the certification of organic products, the income of the farmers will be further improved over the years.Realising the success, the IOFS activities have been extended to two more villages i.e. Umdan Umbathiang and Pynthor in the Umsning Bloick of Meghalaya.

Thus, it can be concluded that OF promotes soil health and provides multiple ecosystem services. IOFS will not only promote organic food production but also reduce dependence on external resources through efficient recycling of on-farm biomass and other resources. It is possible to earn about Rs. 1,45,00/ha (Rs.12000/month) through adoption of IOFS practice. Application of lime, bio-fertilizer, liquid manure along with adoption of time tested ITKs can meet the nutrient requirement of the IOFS farm when on-farm resources are efficiently recycled. There is urgent need for locations specific IOFS models for upscaling in NEH region of India.

References

Bujarbaruah, K.M. 2004. Organic Farming: Opportunities and Challenges in North Eastern February, 2004. ICAR Research Complex for NEH Region, Umiam, Meghalaya.

Das, Anup, Patel, D.P., Kumar, M., Ramkrushna, G.I, Ngachan, S.V.,Layek, Jayanta and Lyngdoh, Mark, 2014. Influence of cropping systems and organic amendments on productivity and soil health at mid altitude of North-East India. *Indian Journal of Agricultural Sciences,* 84(12): 1525–30.

Das Anup, Patel, D.P, Kumar, M, Ramkrishna GI, Mukherjee, A, Layek J, Ngachan, S.V. and Buragohain, J. 2017. Impact of seven years of organic farming on soil and produce quality and crop yields in eastern Himalayas, India. *Agriculture Ecosystem and Environment,* 236: 142-153

Sharma, U.C., Samra, J.S. and Prodhani, M.A. 2006. Land and environmental degradation. In: Soils and their management in north-east India (eds.) Sharma, U.C., Datta, M. and Samra, J.S. ICAR research Complex for NEH region, Umiam, Meghalaya, India, pp. 122-170.

Willer, H. and Yussefi, M. 2005. The world of organic agriculture – statistics and emerging trends, 2005. Bonn, Germany: International Federation of Organic Agriculture Movements (IFOAM), pp. 196.

❑❑❑

Chapter 15

Techniques for Establishment of Tea on Terrace Riser in the Nilgiris

***Anup Das, Subhash Babu, Jayanta Layek, Ramkrishna G.I., Thoi Thoi Devi, Dauni Suiting, Utpal Dey, GS Yadav and S.V. Ngachan**

ICAR - Indian Institute of Soil and Water Conservation, Research Centre, Udhagamandalam-643004. Nilgiris, Tamil Nadu, E-mail: opskhola@gmail.com

Introduction

The Nilgiris is a fragile hilly eco-system bestowed with potentially rich and diversified natural resources but subjected to many land degradation processes, which hinder its development. Bench terracing has been practised as a soil and water conservation measure on the steep hill slopes since time immemorial. Bench terracing consists of a series of platforms along contours SWC at regular intervals separated by vertical drops or risers. In effect these reduce the land slope, intercept the long slope into several small strips and make protected platforms available for conservation farming in hilly area. Bench terraces may be level (table top) or sloping outwards or inwards depending on local conditions like land slope, soil conditions, rainfall in the area, farming practices to be adopted, etc. Inward sloping terraces are suitable for high rainfall areas with deep permeable soils, like that of the Nilgiris.

The Madras State has been one of the earliest to launch the scheme for bench terracing in the Nilgiris district as early as 1951 on an area of 6000 ha. After conducting several experiments on bench terracing over the years, ICAR - Indian Institute of Soil and Water Conservation, (ICAR-IISWC) Research Centre, Udhagamandalam had come out with the type and specifications of bench terraces suitable for the areas receiving high rainfall, possessing steep slopes and where annual crop cultivation is taken up. Bench terraces have been recommended to be formed with 2.5% inward slope and 1% longitudinal slope with risers maintained at 0.5:1 slope when soil is heavy textured and 1:1 slope when soil is light textured.

Bench terracing requires initial heavy investment and proper maintenance thereafter. In the Nilgiris, proper maintenance of terraces is not in vogue leading to improper terrace conditions thereby increasing the erosion hazards and defeating the very purpose for which the terraces were initially formed. Farmers more often than not tend to keep their terraces outward sloping though the benefits of inward sloping terraces have been well-established. This can be attributed to the fact that formation of inward sloping terraces involves considerably higher soil movement than that of outward sloping terraces. Farmers who are skeptical of the high soil disturbance tend to stop with outward sloping terraces. With the passage of time, even terraces which were originally inward sloping had got converted into moderately to steeply outward sloping terraces due to poor maintenance.

There is a common grouse among farmers in Nilgiris that cultivable area gets reduced when riser slope of 0.5:1 or 1:1 is provided. So, more often than not, farmers tend to make the riser as vertical as possible in order to accommodate more crop rows and minimize the loss of cultivable area in the risers, which leads to riser instability. In order to overcome this problem, the area loss due to riser slope, can be converted to additional area available for cultivation, if some perennial crops having economic value are grown on these risers. This not only ensures crop diversity in the same field but also minimizes the risk of crop failure, besides providing additional income to the farmer. ICAR-IISWC Research Centre, Udhagamandalam had evaluated various perennial species for their suitability to be grown on terrace risers during the past several years and found tea to be the most promising alternative. Growing tea on terrace risers in the Nilgiris is a profitable proposition with the added advantage of riser protection since tea possesses good canopy coverage, high crop longevity and also has a good market value. Additional returns through an yield of 5.8 t/ha/yr of green tea leaves can be obtained from the third year onwards, making it a viable option for farmers. However, the efforts made to extend this technology on farmers' fields have met with failure probably due to the improper method of establishing tea on terrace risers. As on now, the practice is to plant tea clones in crow bar holes without any basal application of fertilizer and/or farm yard manure (FYM).

Tea is well known as fertilizer dependent crop and its response to N, P and K has been well-documented (Sen, 1967; Dey, 1971; Rahman, 1977). Wanyoko (1997) reported that tea cannot be produced optimally without fertilizer application. Integrated Soil Fertility Management (the combined use of organic and inorganic fertilizers) is recommended because of its ability to concurrently increase crop yield and soil health (Vanlauwe *et al.*, 2001). Islam *et al.* (2017) observed that tea plant needs large amount of micro and macro nutrients for its growth and that the deficiency of these nutrients could adversely affect the yield and quality. These nutrients are critical and their deficiency and inadequate supply would eventually lead to poor seedlings establishment and performance.

Experimental results have shown that the increase in soil organic carbon by 1% could increase annual yield by about 400-500 kg/ha (Amarathunga and Wijeratne, 2000). Irrigation also increases tea yield by about 50-100%. The yield response to fertigation i.e. irrigation with fertilizer, can be as high as 300% during dry weather (Anonymous, 2001 and 2002). A study undertaken by Bandara (2011) in the low elevation tea growing areas of Sri Lanka confirmed that irrigation can be used as a viable option for improving productivity and to mitigate short term drought effects on tea. He reported that it was crucial to irrigate for establishment of young plants as early growth determines yield potential later. Tea soils should have a high capacity of water retention, but water should not remain stagnant (Willson, 1999). Lack of soil moisture decreases the growth of the branches, and as a result, the leaves become very hard and tend to produce a number of sterile buds, which reduces the yield and quality (Lemmesa, 1996).

As the practice of planting in crow bar holes without application of fertilizers and/or farmyard manure had not lead to satisfactory levels of tea establishment on the risers of terraces in farmers' fields, an experiment was conducted at ICAR-IISWC, Research Centre, Udhagamandalam to find out suitable ways of establishing tea on terrace risers under different bench terrace conditions in the Nilgiris.

Materials and Methods

The experiment was conducted on existing terraces of the Research Farm of ICAR-IISWC Research Centre, Udhagamandalam where general land slope was in the range of 28-30%. Entire experimental area was divided into two portions. In one portion, inward sloping bench terraces (2.5%) and in another portion outward sloping terraces (6-8%) were formed by inducing manual landscaping on the existing terraces. Each of these two types of terraces were further divided into two portions to have two types of irrigation systems for facilitating the establishment of tea and each of these portions had four treatments of soil working and planting, which is the third factor. The experiment was laid out in split plot design with three replications.

Treatments

Factor-1: Types of bench terrace - Two levels

a) Inward sloping bench terraces (2.5% inward slope)
b) Outward sloping bench terraces (6-8% outward slope)

Factor-2: Irrigation methods - Two levels

a) *Jerry* (Local irrigation method)
b) Drip

Factor-3: Soil working and planting method

a) Planting in crow bar hole (Earlier recommended practice)
b) Planting in crow bar hole with application of FYM and fertilizers
c) Planting in pits of 20 cm diameter and 20 cm depth
d) Planting in pits of 20 cm diameter and 20 cm depth with application of FYM and fertilizers.

Following the above design, 48 riser plots were made, with each plot measuring 6 meters in length. Tea clones (B6/61) were planted in two parallel rows on terrace risers with the recommended spacing of 75 X 75 cm during August, 2008. Vegetable cultivation (potato-cabbage-carrot) was taken up uniformly in all the bench terraces. Since December, 2008, irrigation treatments were imposed in the experiment after the withdrawal of monsoon and applied during the dry spell from January-March, every year. Drip irrigation system was installed in 24 plots while *jerry* irrigation was given in the rest. In both methods of irrigation, water was applied every week at the rate of 128 litres per plot containing 16 plants. In drip method of irrigation, the application rate was 4 litres per plant twice a week while in *jerry* method, water was applied once a week for the entire plot. Tea planted on the terrace risers in the experimental site had encountered stem borer attack in the year 2011 and in the process had suffered casualties. Gap filling was done by replanting, wherever the damage had occurred.

Growth parameters of tea like plant height, number of branches per plant, number of leaves per plant and canopy percentage were monitored in all experimental plots (48 Nos.) on yearly basis for four years. Yield of tea was monitored from third year onwards for two years. Soil samples were collected initially from the experimental site at two depths (0-15 and 15-30 cm) and tested for soil pH, EC, organic carbon and status of available N. Soil samples were collected from each riser plot (48 Nos.) at two depths (0-15 and 15-30 cm) four years after planting tea on the terrace risers and analyzed for soil pH, EC, organic carbon and status of available N, to discern the change in soil properties after the imposition of different treatments. Monitoring of soil moisture status was conducted during the dry spell of three months from January- March, every year, the period during which irrigation treatments were imposed.

Results and Discussion

Growth parameters: The growth parameters of tea like plant height, canopy, number of branches per plant and number of leaves, which were monitored over four years after planting on terrace risers are presented in Tables 1 to 4.

Table 1: Average plant height of tea under different treatments (2009-10 to 2012-13)

Treatment	Plant height (cm)			
	2009-10	2010-11	2011-12	2012-13*
Type of bench terrace				
Inward sloping bench terrace	36.1	69.9	83.7	58.4
Outward sloping bench terrace	35.1	73.3	74.6	50.2
CD (P=0.05)	NS	NS	NS	2.6
Method of irrigation				
Drip irrigation	35.9	71.2	78.7	54.2
*Jerry*** method of irrigation	35.3	72.0	79.5	54.4
CD (P=0.05)	NS	NS	NS	NS
Soil working and planting method				
T_1 : Planting in Crow bar hole	30.0	62.9	73.5	50.4
T_2 : T_1 + FYM & Fertilizer	39.9	77.4	84.0	57.1
T_3 :Planting in bigger hole (20 cm dia and 20 cm depth)	31.4	65.1	72.9	49.4
T_4 : T_3 + Application of FYM and Fertilizer	41.1	81.1	86.2	60.2
CD (P=0.05)	1.8	7.84	6.6	4.1

*Replanting had been carried out in 2011 to counter damage due to stem borer attack; **Local method of irrigation in the Nilgiris

It can be observed from Table 1 that plant height of tea planted on terrace risers did not vary significantly with treatments pertaining to terrace system or the method of irrigation adopted during dry spells. However, the treatments which pertained to soil working and planting showed significant difference over all the four years of monitoring. When tea was planted in bigger holes with 20 cm diameter and 20 cm depth along with the application of farmyard manure and fertilizers, higher plant height was observed over the years.

Data on the canopy of tea monitored under different treatments (Table 2) reveal that type of bench terrace and irrigation method did not significantly influence it. However, significant difference could be observed in canopy with respect to soil working and planting method during the first three years after planting. Planting tea in bigger holes coupled with the application of farmyard manure and fertilizers had increased the canopy of tea in comparison with other treatments.

Table 2: Average canopy of tea under different treatments (2009-10 to 2012-13)

Treatment	Canopy (%)			
	2009-10	2010-11	2011-12	2012-13
Type of bench terrace				
Inward sloping bench terrace	13.3	32.4	29.2	52.9
Outward sloping bench terrace	14.4	26.7	33.0	44.7
CD (P=0.05)	NS	NS	NS	NS
Method of irrigation				
Drip irrigation	13.2	29.4	30.7	47.5
*Jerry*** method of irrigation	14.5	29.7	31.4	50.1
CD (P=0.05)	NS	NS	NS	NS
Soil working and planting method				
T_1 :Planting in Crow bar hole	11.6	21.6	24.5	47.1
T_2 : T_1 + FYM and Fertilizer	15.2	35.9	34.5	49.8
T_3 :Planting in bigger hole (20 cm dia and 20 cm depth)	12.4	25.1	25.7	46.8
T_4 : T_3 + Application of FYM and Fertilizer	16.1	35.7	39.6	51.5
CD (P=0.05)	1.0	10.1	5.7	NS

No significant difference could be observed in the average number of branches per plant (Table 3) and number of leaves per plant (Table 4) due to type of bench terrace and method of irrigation during dry spells. Only soil working and planting method caused significant difference in both parameters during the initial two years after planting tea on terrace risers.

Table 3: Average number of branches per plant in tea under different treatments (2009-10 to 2012-13)

Treatment	No. of branches/plant			
	2009-10	2010-11	2011-12	2012-13
Type of bench terrace				
Inward sloping bench terrace	4.3	18.4	34.3	59.6
Outward sloping bench terrace	4.5	15.9	42.7	55.3
CD (P=0.05)	NS	NS	NS	NS
Method of irrigation				
Drip irrigation	4.3	16.2	36.5	57.1
*Jerry*** method of irrigation	4.5	18.1	40.6	57.8
CD (P=0.05)	NS	NS	NS	NS
Soil working and planting method				
T_1 : Planting in Crow bar hole	4.2	13.4	36.4	52.9
T_2 : T_1 + FYM and Fertilizer	4.6	18.9	37.8	60.7
T_3 :Planting in bigger hole (20 cm dia and 20 cm depth)	4.2	14.7	42	53.6
T_4 : T_3 + Application of FYM and Fertilizer	4.8	21.7	37.9	62.6
CD (P=0.05)	0.4	3.64	NS	NS

Table 4: Average number of leaves per plant in tea under different treatments (2009-10 to 2012-13)

Treatment	No. of leaves/plant			
	2009-10	2010-11	2011-12	2012-13
Type of bench terrace				
Inward sloping bench terrace	25.5	114.1	231.9	647.6
Outward sloping bench terrace	31.3	111.9	237.6	503.5
CD (P=0.05)	NS	NS	NS	NS
Method of irrigation				
Drip irrigation	27.8	108.8	227.7	567.1
*Jerry*** method of irrigation	29.1	117.3	241.7	584.0
CD (P=0.05)	NS	NS	NS	NS
Soil working and planting method				
T_1 :Planting in Crow bar hole	23.3	94.27	209.5	512.5
T_2 : T_1 + FYM and Fertilizer	31.0	123.8	259.2	601.1
T_3 :Planting in bigger hole (20 cm dia and 20 cm depth)	25.5	101.4	236.9	561.9
T_4 : T_3 + Application of FYM and Fertilizer	34.0	132.6	233.2	626.5
CD (P=0.05)	2.7	22.6	NS	NS

The growth of tea plants, as monitored by plant height, canopy cover, number of branches per plant and number of leaves per plant, over four years was found to be better, when larger pit size (20 cm dia with 20 cm depth) was adopted with the application of FYM and fertilizers. The results are in line with the work undertaken by Ipinmoroti *et al.* (2009) who reported that use of both organic and inorganic fertilizer materials improved tea plant growth performance indicating that the soil requires the application of fertilizer for tea cropping. It can also be discerned that tea planted on the risers of inward sloping terraces fared better in comparison with that planted on the risers of outward sloping terraces. Growth of tea established with *jerry* system of irrigation during dry spells was better than that with drip irrigation.

Yield: The yield of tea planted on terrace risers in the experimental plot was monitored for two years and presented in Table 5. The type of terrace, irrigation method and soil working and planting methods had not significantly affected the yield of tea. However, it is imperative to note that tea planted in bigger holes with the application of farm-yard manure and fertilizers had attained higher yield than other soil working and planting methods.

Table 5: Average yield of tea under different treatments (2011-12 and 2012-13)

Treatment	Tea yield (kg/ha)	
	2011-12	**2012-13**
Type of bench terrace		
Inward sloping bench terrace	438.1	962.9
Outward sloping bench terrace	411.5	977.1
CD (P=0.05)	NS	NS
Method of irrigation		
Drip irrigation	479.7	1034.3
*Jerry*** method of irrigation	369.9	902.9
CD (P=0.05)	NS	NS
Soil working and planting method		
T_1 :Planting in Crow bar hole	430.1	800.0
T_2 : T_1 + FYM & Fertilizer	432.5	1054.3
T_3 :Planting in bigger hole (20 cm dia and 20 cm depth)	396.4	974.3
T_4 : T_3 + Application of FYM & Fertilizer	440.1	1057.1
CD (P=0.05)	NS	NS

Soil fertility parameters: Table 6 indicates the initial status of soil properties at the experimental site and Tables 7 and 8 indicate the status of soil properties at two soil depths (0-15 cm and 15-30 cm) under different treatments, four years after planting tea on terrace risers. It can be observed from Table 6 that the soil of the experimental site is non-saline and acidic in nature with high organic carbon content and medium N.

Table 6: Initial status of soil properties at experimental site

Depth (cm)	pH	EC (dS/m)	OC (%)	Available N (kg/ha)
0-15	4.53	0.11	1.57	329.9
15-30	4.73	0.05	1.49	296.6

Table 7: Status of soil properties (0-15 cm) under different treatments four years after planting tea on terrace riser

Treatments	pH	EC (dS/m)	OC (%)	Available N (kg/ha)
Type of bench terrace				
Inward sloping bench terrace	5.04	0.06	2.60	434
Outward sloping bench terrace	5.01	0.05	2.70	376
CD (P=0.05)	NS	NS	NS	NS
Method of irrigation				
Drip irrigation	5.02	0.06	2.63	401
Jerry method of irrigation	5.03	0.05	2.66	410
CD (P=0.05)	NS	NS	NS	NS
Soil working and planting method				
T_1 :Planting in Crow bar hole	5.01	0.05	2.65	434
T_2 : T_1 + FYM and Fertilizer	5.04	0.05	2.60	336
T_3 :Planting in bigger hole (20 cm dia and 20 cm depth)	5.02	0.05	2.64	410
T_4 : T_3 + Application of FYM and Fertilizer	5.03	0.07	2.69	441
CD (P=0.05)	NS	NS	NS	NS

Table 8: Status of soil properties (15-30 cm) under different treatments four years after planting tea on terrace riser

Treatments	pH	EC(dS/m)	OC (%)	Available N (kg/ha)
Type of bench terrace				
Inward sloping bench terrace	5.07	0.03	2.13	354
Outward sloping bench terrace	5.01	0.03	2.02	474
CD (P=0.05)	NS	NS	NS	NS
Method of irrigation				
Drip irrigation	5.05	0.03	2.14	420
Jerry method of irrigation	5.03	0.03	2.01	408
CD (P=0.05)	NS	NS	NS	NS
Soil working and planting method				
T_1 :Planting in Crow bar hole	5.01	0.04	2.11	403
T_2 : T_1 + FYM and Fertilizer	5.08	0.03	2.24	413
T_3 :Planting in bigger hole (20 cm dia and 20 cm depth)	5.01	0.03	1.91	399
T_4 : T_3 + Application of FYM and Fertilizer	5.06	0.03	2.03	441
CD (P=0.05)	NS	NS	NS	NS

Analysis of soil samples collected four years after tea planting did not indicate any significant effect of the treatments on soil fertility parameters. However, it could be observed that after four years of planting tea on terrace risers, soil pH had increased, EC had reduced and soil organic content and available N had increased across the different treatments, when compared with the initial status of soil properties at the experimental site.

Soil moisture: The average soil moisture content monitored under the different treatments during the dry spells of a year (2009-10 to 2012-13) is presented in Table 9. It can be observed from this table that in three out of four years, soil moisture content was significantly higher during the dry spells in the treatment, where tea was planted in bigger hole (20 cm dia) with the application of farmyard manure and fertilizers over the standard planting practice in crow bar hole (15 cm dia). Type of bench terraces and irrigation methods did not have any significant effect on soil moisture status, though outward sloping terraces and drip irrigation method had higher moisture content than inward sloping terraces and *jerry* method of irrigation during most of the years of monitoring.

Table 9: Status of soil moisture during dry spell under different treatments

Treatments	Average soil moisture during dry spell (% by weight)			
	2009-10	2010-11	2011-12	2012-13
Type of bench terraces				
Inward sloping bench terrace	26.1	25.3	20.7	15.5
Outward sloping bench terrace	26.5	25.7	21.1	15.2
CD (P=0.05)	NS	NS	NS	NS
Method of irrigation				
Drip irrigation	26.4	25.6	21.3	15.0
Jerry method of irrigation	26.2	25. 5	20.6	15.6
CD (P=0.05)	NS	NS	NS	NS
Soil working and planting method				
T_1 :Planting in Crow bar hole	24.7	25.1	20.2	15.3
T_2 : T_1 + FYM and Fertilizer	26.4	24.3	20.2	15.5
T_3 :Planting in bigger hole (20 cm dia and 20 cm depth)	26.6	25.8	21.7	14.9
T_4 : T_3 + Application of FYM and Fertilizer	27.5	26.8	21.4	15.6
CD (P=0.05)	0.8	1.25	1.01	NS

Conclusions

From the experiment carried out to find suitable techniques for establishing tea on the risers of bench terraces in the Nilgiris, it was established that the type of bench terrace *viz.* inward or outward sloping had no significant effect on the growth parameters of tea, which were planted on the terrace risers. Also, for initial establishment of tea on the risers, two methods of irrigation were tested to safeguard the plants against moisture stress during dry spells of a year. Both drip and *jerry* methods of irrigation did not significantly influence the growth parameters of tea. However, out of the four soil working and planting methods followed, it was found that planting tea in bigger holes (20 cm dia) along with the application of farmyard manure and fertilizers aided in better growth of tea as witnessed by higher plant height, canopy, number of branches and number of leaves in a tea plant. Soil properties like pH, EC, organic matter content and available N underwent no significant change due to the different treatments imposed. Also, soil moisture status during dry spells of a year was not significantly influenced by the imposed treatments. It can be concluded that for establishing tea on terrace risers, irrespective of the type of terrace and the method of irrigation provided during the initial years of establishment, planting the tea clones in larger sized holes than the conventional crowbar sized holes along with the application of farm yard manure and fertilizers should be followed. This will result in better growth parameters and yield of tea and serve as a viable option for farmers, safeguarding their landholdings against erosion and providing them with a sustainable source of income.

References

Amarathunga, M.K.S.L.D. and Wijeratne, M.A. 2000. Some of the ecological factors affecting sustainable productivity of clonal tea *Proceedings of the 200th Experiment and Extension Forum.* The Tea Research Institute of Sri Lanka, St Coomb's Estate, Talawakelle, 515p.

Anonymous, 2001 & 2002. Reports of the Agronomy Division, Tea Research Institute of Sri Lanka, St Coomb's Estate, Talawakelle.

Bandara, Shaymantha Nelum, 2011. Thesis on Agronomy of irrigated tea in low elevation growing areas of Sri Lanka submitted to School of Agriculture, Food and Wine, The University of Adelaide.

Dey, S.K. 1971. Potash as a limiting factor. *Two and A Bud,* 18: 34-47.

Ipinmoroti, R.R., Iremiren, G.O., Lubamiwa, O.O., Fademi, A.O. and Aigbekaen, E.O. 2009. Effect of inorganic and organic based fertilizer on the growth performance of tea and cost implication in Kusuku, Nigeria. Pro. 9th African Crop Science, Conference Proceedings, Cape Town, South Africa, PP. 247-250.

Islam, Shamsul; Hamid, F.S.; Amin, Khawar; Sumreen, Sonia; Qamar-uz-Zaman, Khan, Noorullah; Khan, Abbas; Shah; Basharat Hussain 2017. Effect of organic fertilizer on the growth of tea (Camellia sinensis L.). *International Journal of Sciences: Basic and Applied Research* 36(8): 1-9.

Rahman, F. 1977. Nitrogen manuring in mature tea. *Occasional Scientific Paper No.* 11, TRA, Jorhat, Assam.

Sen, A.R. 1967. Responses of mature tea to phosphate and potash in North-East India. *Expt. Agric.,* 3: 55-62.

Vanlauwe, B., Wendt, J., and Diels, J. 2001. *Combined application of organic matter and fertilizer*. In G. Tian, F. Ishida & J.D.H. Keating (Eds.), *Sustaining Soil Fertility in West Africa* (SSA Special Publication No. 58). Soil Science Society of America, Madison, USA.

Wanyoko, J.K. 1997. Advances in fertilizer use in Kenyan tea industry. *Tea,* 18: 123-131.

Lemmesa, F. 1996. *Tea production and management* (pp. 1–14). Teaching handout, Department of Plant Sciences, Jimma College of Agriculture, Jimma, Ethiopia.

Willson, K.C. 1999. *Coffee, cocoa, and tea. Crop production science in horticulture series 8*. Cambridge, U.K.: CABI Publishing.

□□□

Chapter 16

Promotion of Artificial Rainwater Harvesting Pond based Integrated Farming System for Tribal *Jhumias* – A New Initiative

Sanjay Kumar Ray, Santosh Kumar Baishya, P. Chowdhury and D.J. Rajkhowa

Krishi Vigyan Kendra,Wokha, ICAR Research Complex for NEH Region, Nagaland Centre, Jharnapani-797106, E-mail: dr.sanjaykumarray@yahoo.co.in

ABSTRACT

Acute scarcity of water during post-rainy season is a common phenomenon in sloppy *jhum* land of Wokha district in Nagaland to carry out different agricultural activities. It is due to the non-availability of efficient water harvesting facility. Farmers are mainly depending on the *jhum* farming system with meager profit. To make *jhum* cultivation into a profitable and sustainable practices, KVK Wokha initiated a new intervention through establishing artificial rainwater harvesting pond based integrated approach of agriculture, horticulture, livestock and fish components on community basis. Three numbers of artificial rainwater harvesting (size; 35m x 10m x 2.5m) embankment type ponds were dug out at Longsa, Riphyim and Yanthamo villages of the district, where 1000 micron UV black colour HDPE sheet was laid down on the pond to store 8.75 lakh litres of water. Ponds were preferably constructed just below the top-hill for efficient utilization of the harvested water for different agriculture and allied components. Tree bean as MPTs were planted in the top of the hills, fruits crops like litchi, Assam lemon, orange and papaya were planted in the mid-hills and banana in the bottom of the hill. Seasonal agriculture crops and vegetables were sown as inter-crop in between the fruits and plantation crops. After demonstration, it was found that the technology was adopted 100% by 28 *jhumias* farmers on community basis. The adoption of technology showed an increased yield of cereal (20%), vegetables (303%), spices (32%), fruits (10%) and reduction of livestock feed cost (34%). In addition, integration of pig cum fish and fruits components fetched an additional average income of ' 1.3 lakhs per unit for the farmers. It was also recorded that the net returns of ' 1.78 lakhs to 2.09 lakhs with BC ratio of 2.97 to 4.27 was accrued from the artificial rainwater harvesting pond based IFS intervention. Finally, it could be concluded that artificial rainwater harvesting pond based IFS model could be a sustainable and economically viable intervention for the tribal *jhumias* farmers.

Keywords: Artificial Pond, Integrated Farming System Jhum Cultivation

Introduction

Shifting or *jhum* cultivation is the most traditional and principal land-use system that is pre-dominantly followed by the chunk of the farmers in Nagaland. Presently, around 1.9 lakh tribal families have been involved in *jhum* farming with an area of 1.24 lakh ha in the state. Among all the districts of Nagaland, the Wokha district has occupied the highest *jhum* cultivating area of 11,670 ha (Statistics of Govt. of Nagaland, 2011). The traditional *jhum* cultivation practice is mostly confined to the steep hill slopes, which is directly associated with huge loss of soil (30 to 170 t $ha^{-1}yr^{-1}$) (Saha *et al.*, 2012), essential plant nutrients, soil flora and fauna that leads the land into fallow and slowly it becomes degraded. During past few decades, the fallow period of the *jhum* cycle has further shortened from 10-15 years to 3-4 years due to the increase of population pressure and food grain demand and that is fetching threats to natural resources and limiting the crop yield. Because of shortening of jhum cycle, now *jhum* cultivation system is considered as a non-productive and unsustainable system.

Agriculture and allied departments have been working towards improvement of *jhum* cultivation system along with sporadic plantation of MPTs and fruits trees for the reforestation of degraded land. The major focus has been given on soil and water conservation measures by executing both the mechanical and biological conservation measures on hill slopes. However, very less attempts have been made to move forwards towards the integrated farming approach entailing the agriculture, horticulture, soil conservation, forestry, fishery, animal husbandry and other allied enterprises; the reason of this being be due to the inadequate water harvesting and storage facilities prevailing in the district. Conventional farm ponds in *jhum* hilly tracks are seasonal and frequently exposed to the potential losses like infiltration, percolation, seepage flow and evaporation loss. Besides these, the traditional methods for construction of farm pond using of concrete, bricks, stone-slabs, etc. are very expensive and difficult to take up for the poor tribal Jhumias. Acute water scarcity during the post-rainy season (November to April) of the year was identified as the main constraint to the jhumias. The harvesting of rainwater and its efficient utilization as an integrated farming system approaches holds a special position rather than the mono-cultural approaches to enhance the water productivity and secure livelihoods in a sustainable manner for rural poor.

Therefore, keeping the above in backdrop, an attempt was made to set an example to transform the existing primitive type of *jhum* farming system into a sustainable integrated farming system model through the establishment of a artificial pond based model to integrate agriculture, horticulture, livestock and other components within the bio-physical and socio-economic environment of the Jhumias of Wokha district.

Materials and Methods

Location and climate of study area

The present study was conducted at three different locations at Yanthamo (site I), Longsa (site II) and Riphyim (site III) villages of Wokha district. The mean annual rainfall for the last five years was recorded 1996 mm with more than 88% occurring during the period from April to October. The average monthly temperature during the year was recorded to be varied widely between 7.4%C to 30.1%C. Wind speed was recorded the lowest (2.7 km h^{-1}) in October and the highest (4.8 km h^{-1}) in March.

The initial physical and chemical properties of the soil varied among the three sites but homogeneous within the sites in texture, pH, SOC, available N, P_2O_5 and K_2O. Soils of the site I, II and III had sandy loam, silty loam and clay loam in texture (Table 1), respectively. According to rating limits of soil test values, soil was strongly acidic in nature, with high organic carbon, low in available nitrogen, low to medium in available phosphorus and high in available potassium content.

Training

A four days training programme was conducted at KVK, Wokha for 19 participants of the three selected sites for capacity building of the farmers on the community basis rainwater harvesting and its efficient utilization through integrated farming system (IFS) approach for sustainable agriculture.

Establishment of Artificial Rainwater Harvesting Pond

The artificial rainwater harvesting ponds were constructed near the top of the hill location on the side of the mound. The artificial embankment type of rainwater harvesting (size; 35m x 10m x 2.5m) pond was constructed just below the top-hill for the efficient utilization of the harvested water through gravitational force, where the water harvesting capacity was 8.75 lakh litres. The 1000 micron UV black colour HDPE sheet was laid down on pond to manage the seepage water loss. Pond were filled up with water from the nearby seasonal springs.

Components of IFS

Altogether 3 numbers of artificial rainwater harvesting pond based models were established involving the artificial pond, crops, MPTS, poultry, pig, fish, low-cost poly house, mushroom and vermi-composting. MPTs like Tree bean and Alder were planted in the top of the hills, fruits crops like Litchi, Assam lemon, Papaya and Plum planted in the middle hills and banana in bottom hill. Seasonal cereal (paddy), oilseed (soyabean), pulse (ricebean), fodder (Maize-HQPM1, cassava, sweet potato), vegetables (colocasia, french beans, garden pea and cabbage), spices (ginger) and were sown as inter-crop in between the fruits and plantation crops. Cucumber and Naga king chilli were taken up as off-season crops, under the low-cost poly house to obtain additional farm income. Oyster mushroom and vermi-composting were also introduced to recycle the crop residue and weed biomass, respectively. Pig and poultry were selected under livestock component. Pig house was constructed nearby the artificial pond. Before release the fish species (Silver carp, Caltla and Grass carp) a 10-15 cm layer of a mixture of surface soil and manures were deposited into the artificial pond for better plankton growth for fish feeds. The site specific components details are given in Table 2.

Table 1: Selected sites location and soil characteristics details

Sites	Location Latitude	Location Longitude	Altitude (m)	Textural Class	pH	SOC (%)	Available N	Available P_2O_5	Available K_2O
Yanthamo (I)	26?03.753'N	094?18.195' E	973	Sandy Loam	5.33	1.895	230.25	21.23	206.98
Longsa (II)	26?04.029'N	094?14.936' E	1174	Silty Loam	4.09	2.560	178.50	33.05	239.81
Riphyim (III)	26?12.282'N	094?15.748' E	536	Clay Loam	3.89	1.250	210.11	18.70	248.64

Table 2: Site specific different component details

Sites/ Components	WHP (cubic m)	MPTs (ha)	Crops						Livestock		Fishery			Others	
			Horticulture			Agriculture (ha)			Pig (nos)	Poultry				Vemi compost	Mushroom
			Fruits (ha)	Vegetables (ha)	Spices (ha)	Poly house (sqm)	Cereal	Oilseed	Pulse	Fodder					
Yanthamo (I)	875	0.4	0.7	0.7	0.6	60	0.2	0.4	0.2	0.2	06	100	3000	1 unit	120 bed
Longsa (II)	875	0.2	0.8	0.3	0.2	-	0.1	-	0.1	0.2	05	-	2000	1 unit	-
Riphyim (III)	875	0.6	1.2	0.8	0.6	60	0.3	0.6	0.1	0.4	11	300	3200	1 unit	230 bed

Soil conservation measures likes contour bunding and terracing were undertaken for middle and gentle slope areas, whereas the half moon terracing was made for fruits crops with diameter of 1.0-1.5 m. All the seasonal crops were grown with the standard package and practices which includes proper seed rate, spacing, water management, weeding, and disease pest management, respectively. However, lime, manures and fertilizers were applied as per the initial soil pH, SOC and available NPK test results.

Water Productivity Estimation

Water productivity was calculated to estimate average amount of output per unit of water applied on a field. Water productivity (WP) was simply calculated by using the equation of:

WP (kg per m^3) = Output (kg per ha)/Water Applied (m^3 per ha) (Wichelns, 2014).

Statistical Analysis

Data on various parameters including management of crops and animals, water necessity, labour involvement, expenditure requirement, total production, and net outcomes was recorded. The adjacent conventional *jhum* fields' production was considered as control plot for each site.

Results and Discussion

Water Harvesting

Water is an essential and integral part for establishment of integrated farming system. It was observed that the capacity of the installed artificial rainwater harvesting pond was about 8.5 to 8.7 lakh litre of water. It was successfully filled up by the end of May at all the three different sites. During the *kharif* season, water only utilized for the drinking purpose of poultry and piggery and also for the production of mushroom, off-season vegetables and vermi-composting. No irrigation was required to grow the seasonal agricultural and horticultural crops during the *kharif* season, but during the *rabi* season harvested water was effectively utilized to grow the winter vegetables and also as life saving irrigations for fruits crop. Fish was also successfully reared into these artificial pond to increase the water productivity.

Fruit Production

Fruit is the highly preferred component for the hilly farmers of the district and the high soil potassium content also support to grow the fruits crops. Earlier it was observed that the farmers were growing fruits in unscientific manner without maintaining its proper spacing, training and pruning operation, and irrigation, fertilization, disease-pest and weed management correctly. Supplement of quality planting materials, irrigation, and integrated nutrient management practices increased the yield by 7.5%, 12%, 12.3% and 8.3% for litchi, plum, banana, and papaya fruits yield, respectively. Assam lemon has yet not started fruiting. The total gross return from this horticultural component was ' 16,200/- against its cost of cultivation of ' 630/- and fetched the net profit of ' 15,750/- (Data not shown).

Vegetable production

The vegetables component occupied an area of 1.8 ha for all the three sites and the vegetables were grown just below the artificial pond during both the *kharif* and *rabi* season. Several vegetables including colocasia, french bean, cabbage and garden pea were raised in this system. The garden pea and french bean were performed better in terms of survivability and growth during the *rabi* season. Introduction of seasonal vegetable cultivation helped to get regular income throughout the year. On an average yield of vegetable was increased by 172 %, 261%, 327% 449% and 312% for colocasia, *kharif* french bean, cabbage, garden pea and winter french bean, respectively. And these yield enhancements were due to the supplement of quality seeds, adequate fertilizers, proper irrigation and better management practices (Table 4). Vegetables production had increased average gross income which ranges from Rs. 16,800, to 51,000/-.

Table 3: Fruit production (kg) and gross returns

Sites/Fruits	**Litchi**			**Plum**			**Banana**			**Papaya**			**Total**
	P	YI	GR	P	YI	GR	P	YI	GR	P	YI	GR	GR(`)
Yanthamo (I)	10	7.6	400	-	-	-	500	13	1,500	120	9	2,400	4,300
Longsa (II)	-	-	-	300	12	6000	30	10	900	20	8	400	7,300
Riphyim (III)	10	7.4	400	-	-	-	120	14	3,600	30	8	600	4,600
Mean	10	7.5	400	300	12	6000		12.3			8.3		

P= Production (kg), YI= Yield Increase (%), GR= Gross Returns (')

Table 4: Vegetables production (kg) status

Sites/ Vegetables	*Pre-kharif & kharif*						*Rabi* season									Total GR (`)
	Colocasia			French bean			Cabbage			Garden Pea			French bean			
	P	YI	GR	P	YI	GR	P	YI	GR	P	YI	GR	P	YI	GR	
Yanthamo (I)	1700	194	17,000	390	255	7,800	260	302	5,200	425	462	17,000	200	306	4,000	51,000
Longsa (II)	400	159	8,000	100	215	2,000	-	-	-	120	440	4,800	100	317	2,000	16,800
Riphyim (III)	900	164	9,000	250	312	5,000	115	352	2,300	240	444	9,600	140	314	2,800	28,700
Mean		172			261			327			449			312		

P= Production (kg), YI= Yield Increase (%), GR= Gross Returns (')

Spices Production

In the Wokha district, farmers usually grow spices viz., ginger, king chilli and chilli in *jhum* areas for their selves consumption as well as commercial purpose. Earlier the yield of these spices observed poor due to the use of local cultivars and poor management practices. But in IFS model, adoption of good quality planting materials and scientific management practices had increased the ginger yield by 32% as compared to *jhum* yield which might be due to better availability of nutrients at crucial growth period.

Seasonal Cereal, Pulse and Oilseeds Production

In *jhum* cultivation system, farmers usually grow paddy as cereal, soyabean as oilseed and rice bean and rajma as pulse crop. Among the agricultural crops rice bean gave the good performance because of the varietal replacement but the average yield enhancement of rice bean was comparatively less by 8.7%. However, the gross return was relatively high in sites I & III for cereal, pulse and oilseed of ' 34,275/- and ' 43,600/-, respectively. The productivity of these crops under farming system increased significantly due to the varietal replacement, liming and scientific management of soil and crops.

Table 5: Fish production and returns in Artificial Pond based Farming System.

Sites/Parameter	Area (ha)	Yield (kg)	Cost (`)	Gross Returns (`)	Net Returns	B:C ratio
Yanthamo (I)	0.6	6,500	32,000	1,30,000	98000	3.06
Longsa (II)	0.2	1,500	11,200	30,000	18800	1.68
Riphyim (III)	1.0	10,300	56,400	2,06,000	149600	2.65
Mean		6,100	33,200	1,22,000	88,800	

P= Production (kg), YI= Yield Increase (%), GR= Gross Returns (‘)

Table 6: Production of cereal, pulse and oilseed

Sites	Cereal (Paddy)			Oilseed (Soybean)			Pulse (Rice bean)			Total
	P	YI	GR	P	YI	GR	P	YI	GR	GR (`)
Yanthamo (I)	620	34	9,300	824	43	20,600	125	7	4,375	34,275
Longsa (II)	270	27	4,050	-	-	-	70	7.9	2,450	6,500
Riphyim (III)	850	29	12,750	1080	34	27,000	110	12	3,850	43,600
Mean		30			38.5			8.7		

P= Production (kg), YI= Yield Increase (%), GR= Gross Returns (‘)

Cultivation of Fodder Crops and Feed Supplement Scenario

Most of tribal farmers used to provide inferior quality feed to pig because of high cost of computed pig feed. Therefore, emphasis has been given towards the cultivation of fodder crops like maize (HQPM1), cassava, sweet potato and colocasia in the farming system models. It was found that among all these fodder crops, cassava and sweet potato performed better in terms of survivability and growth as the local cultivars were introduced. Productivity of the HQPM 1 was relatively less as compared to its potential capacity and its productivity ranged from 3.8 to 5.26 t/ha. Interestingly, it was observed that the production of these fodder crops supplemented the feed requirement of pig feed by 30.3% which reduced the feeding cost by 34%. At the end, the farmers were started adopting the quality feeding system by supplementing both the commercial feed and on farm produced feed to fed to their pigs and observed better growth.

Pig Rearing

Pig rearing is one of the most valued enterprise for Naga tribes. Local pig breeds are usually smaller in size and less productive. Therefore, under artificial pond based farming system model, the cross-breed was introduced for better productive and reproductive performance. It was observed that the average body weight gain of pigs was about 78 kg for male 71 kg for female after 10 months under this input management system. Pigs were fed daily with commercial feed along with fresh and dry cassava, sweet potato leaves, tubers and local grasses. Out of the total 22 pigs, 6 were males and the remaining 16 pigs were females. Out of the 16 number of pig, 11 pigs had conceived and other yet to be conceived. The major quantities of pig manure were diverted to pond and remaining dung was used to prepare the vermicompost.

Table 7: Fodder production and feed supplement to pig

Sites	Maize (HQPM)			Cassava (Local)			Sweet Potato			Total FS	Feed Cost Reduction (%)
	A	P	FS	A	Production	FS	A	Production	FS		
Yanthamo (I)	0.1	490	5	0.07	1,610	11	0.02	440	4	20	24
Longsa (II)	0.1	526	9	0.10	1,950	13	-	-	-	22	25
Riphyim (III)	0.1	388	4	0.25	7,500	38	0.05	980	7	49	52
Mean	0.1	468	6	0.14	3687	20.7	0.04	710	5.5	30	34

A= Area (ha), P= Production (kg), FS= Feeding Supplement (%)

Table 8: Growth performance of pig under Integrated Farming System

Sites	Number of piglets			Mean Body wt (kg) after 5 months		Mean Body wt (kg)after 10 months		Approx. Present value (`)
	T	M	F	M	F	M	F	
Yanthamo (I)	06	02	04	36.5	32.3	79.5	69.5	34,650
Longsa (II)	05	01	04	34.3	30.5	75.5	70.4	54,712
Riphyim (III)	11	03	08	34.8	29.7	78.4	72.3	1,24,327
Mean	7.3	02	5.3	35.2	30.83	78	71	71,229

Backyard Poultry Rearing and Egg Production

Rearing of poultry is economical only when proper care had been taken otherwise; it is a very much risky enterprise due to frequent mortality and breakout of the severe diseases had experienced by the local farmers. To minimize the mortality of the poultry birds, one month old vanaraja and gramapriya chicks were supplemented to the farming system beneficiary after giving proper treatment and vaccination. The low cost poultry unit was made (sized 100-140 sqft) to rear about 100-120 numbers of birds for both, the meat and egg purpose. The majority of the male chicks were sold at attaining the body weight (10-12 weeks old) of 1.5-2.0 kg and female birds were kept for rearing of egg production. The egg purpose chicks were kept with male: female ratio of 1:8. The litter materials of poultry unit were used as manure for vegetable production. It was observed that in integrated farming system model farmers got ' 16,020 to ' 36,225 as gross returns from the poultry enterprises.

Fish Production

Integrated fish cum pig is the very much preferred farming system for the tribal farmers as majority of the people consumes fish. Before releasing the fish into artificial rainwater harvesting pond a quantity of soil and organic manures were applied for good plankton growth for fish feeding. The fish species of silver carp, catla and grass carp were released with a ratio of 2:3:5, when the water appears green in colour to rear the fish into the ponds. The partial harvesting was made at every one month interval after attaining the average fish size of 200-250 g size and sold at marketable price. Besides, the fish rearing, harvested water was also utilized as life saving irrigation for production of winter vegetables and other purpose for income generating enterprises. Harvested pond water also helped to increase the cropping intensity, productivity and profitability of the models significantly. Fish component accrued an average gross returns of Rs. 52,301/-.

Table 9: Egg and meat production of vanaraja and gramapriya birds

Sites	No. Chicks Distributed	Chicks Sold (Male)			No. of Chicks for Egg Production			Egg Laying		Total Return (`)
		Numbers	Kg	Amount	Total	Female	Male	Number	Amount (`)	
Yanthamo (I)	100	38	74	13,320	62	62	10	450	2,700	16,020
Riphyim (III)	300	124	256	46,080	176	147	29	1825	10,950	57,030
Mean	150	81	165	29,700	119	104.5	19.5	1,137.5	6,825	36,525

Table 10: Fish production and returns in Artificial Pond based Farming System

Sites	No. Fingerlings Released	1st Partial Harvesting			2nd Partial Harvesting			Final Harvesting			Gross Returns (`)
		Number	Kg	Returns (`)	Number	Kg	Returns (`)	Number	Kg	Returns (`)	
Yanthamo (I)	3000	700	105	5,250	1200	312	18,720	770	315.7	25,256	49,226
Longsa (II)	2000	400	64.8	3,240	400	136	8,160	810	502.2	40,176	51,576
Riphyim (III)	3200	1200	174	8,700	1000	270	16,200	780	390	31,200	56,100
Mean	2,733	767	115	5,730	867	239	14,360	787	403	32,211	52,301

Mushroom Production

Two mushroom production unit was also included in the farming system by making a small hut with size of 60 to 100 m^2. Paddy straw and maize stalk were used to raise mushroom by making bamboo racks into the shed. Proper humidity (75-80%) was maintained into the hut during the crop season by sprinkling water by using sprayer over the walls of hut and over bags. Mushroom was raised & harvested at three phases in a year with 120 and 230 number of bag in each cycle for site I & Site III, respectively. During the winter season, growing of mushroom was avoided due to low temperature. It was observed that on an average 913.5 kg mushroom harvested and sold to ' 1, 37, 025.

Vermicomposting

Under artificial pond based model, three numbers of low-cost portable vermicompost units were installed for efficient utilization and recycling of crop residues and weed biomass through vermicomposting techniques. Waste product of different components in farming system and recycling of by-products/farm waste has been practiced. The total biomass produced in existing model from weeds/ crops was recycled by producing as vermicompost or part of the biomass (paddy straw) was used for mushroom production and as a mulching material for winter crops growth. This practice helped in nutrient content of plant wastes increased manifolds after recycling it into vermicompost and minimize expenditures in the cost of production. Earlier, farmers required to burn to re-use the crop residue and weed biomass during the *jhum* farming system and that was because of scarcity of sufficient water for recycling properly through composting or vermicomposting techniques.

Table 11: Mushroom production and gross returns

Sites	Bag storage capacity (nos.)	Phase I (Feb-April)		Phase II (May to July)		Phase III (Aug to October)		Yearly Harvest (Kg)	Gross Returns (`)
		Harvest (Kg)	Returns (`)	Harvest (Kg)	Returns (`)	Harvest (Kg)	Returns (`)		
Yanthamo (I)	120	174	26,100	225	33,750	245	36,750	644	96,600
Riphyim (III)	230	310	46,500	423	63,450	450	67,500	1,183	1,77,450
Mean	175	242	36,300	324	48,600	347.5	52,125	913.5	1,37,025

Table 12: Mushroom production and gross returns

Sites	Bag storage capacity (nos.)	Phase I (Feb-April)		Phase II (May to July)		Phase III (Aug to October)		Yearly Harvest (Kg)	Gross Returns (`)
		Harvest (Kg)	Returns (`)	Harvest (Kg)	Returns (`)	Harvest (Kg)	Returns (`)		
Yanthamo (I)	120	174	26,100	225	33,750	245	36,750	644	96,600
Riphyim (III)	230	310	46,500	423	63,450	450	67,500	1,183	1,77,450
Mean	175	242	36,300	324	48,600	347.5	52,125	913.5	1,37,025

Off-Season Cultivation

Cold climatic condition and water scarcity was identified as the major constrains for the production of year round vegetables. In the farming system models, plenty of available water helps to grow off-season cucumber and king chilli under low-cost polyhouse successfully. The low-cost polyhouse creates micro-climatic environment and that help to grow off-season cucumber and *Naga* king chilli. In this cultivation system, local cucumber sown in the month of January and harvested by the month of March, whereas, *Naga* king chilli was also transplanted by the month of February as inter-crop with cucumber and harvested upto December month. In addition to that nursery for winter crops was also grown under poly house successfully. It was observed that the produce grown under polyhouse was least affected by disease and pest attack. Off-season vegetables cultivation in an area of 60sqm gave the higher gross returns of ' 5800/- to ' 6,600/- due to the high market demand.

Employment generation scenario

Artificial pond based farming system model was providing an abundant scope for employment generation. The cultivation of crops likes cereal, pulse, oilseeds, fodder, seasonal vegetables, off-seasonal vegetables, spices and fruits cultivation required regular farm labours. Out of the different components, crops component created the highest employment opportunities (170 man-days). Besides crops components, integration of fish-cum-piggery, poultry rearing, mushroom production and vermicompost increased the employment opportunities from 196 man-days to 350 man-days/year (Table 14). Most importantly diversified integrated farming practices created some direct and indirect employment opportunities to the beneficiary. Moreover, it generates employment round the year in their farm.

Table 13: Off-season vegetables production and gross returns under polyhouse

Sites	Cucumber			Naga king chilli			Total Gross Returns (`)
	Area (sqm)	Production (kg)	Returns (`)	Area (sqm)	Production (kg)	Returns (`)	
Yanthamo (I)	10x6	50	2,000	10x6	19	3,800	5,800
Riphyim (III)	10x6	60	2,400	10x6	21	4,200	6,600
Mean	60sqm	55	2,200	60sqm	20	4,000	6,200

Table 14: Employment generation (man-days*/year) through artificial pond based farming system.

Sites/ Components	Component-wise man-days requirement in a year											Fishery	Other Enterprise		Total Required Man-days
	Agricultural crops				Horticultural Crops				Total Man days	Livestock					
	Cereal	Oilseed	Pulse	Fodder	Fruits	Vegetables	Spices	Poly house		Piggery	Poultry		Mushroom	Vermi-compost	
Yanthamo (I)	28	40	12	20	15	46	30	25	216	40	40	20	30	26	346
Longla (II)	14	-	6	20	18	22	10	-	90	35		20	30	26	175
Riphyim (III)	42	60	6	40	29	51	30	25	283	75	120	20	30	26	528
Mean	28	50	8	26.7	20.7	39.7	23	16.7	196	50	80	20	30	26	350

Conclusion

Finally, it could be concluded that the artificial pond based integrated farming system could enhance the farm productivity and income of the tribal jhumias through efficient utilisation of water and other local resources. Moreover, this farming system could reduce the risk of soil and environmental degradation, and make the soils more productive. Rapid dissemination and large scale implementation of the artificial pond based technology in the farmers' field will definitely bring phenomenal changes in the agriculture sector of the tribal farming community.

References

Anonymous, 2013.http://nagaland.nic.in/Notification/Microsoft%20 Word%20-20Integrated%20agriculture%20paper% 20190112_ Nagaland.pdf. Accessed 10th January 2017.

Behera U.K., Jha K.P., Mahapatra I.C., 2004. Integrated management of available resources of the small and marginal farmers for generation of income and employment in eastern India. *Crop Research,* 27(1): 83-89.

Saha R., Chaudhary R.S., Somasundaram J. 2012. Soil health management under hill ecosystems of North Eastern, India. *Applied and Environmental Soil Science,* Article ID 696174, 9 pages http://dx.doi.org/10.1155/2012/696174.

Wichelns D., 2014. Water productivity: Not a helpful indicator of farm level optimization. Global water forum www.globalwaterforum.org/2014. (Accessed on 10th January, 2017).

□□□

Chapter 17

Hill Slope Management for Soil and Water Conservation under New Tea Plantation in Nilgiris of South India

D. C. Sahoo[1], M. Madhu[1], S. Santhana Bosu[2] and O. P. S. Khola[3]

[1] ICAR-Indian Institute of Soil and Water Conservation, Research Center, Sunabeda-763002, Odisha, [2] Agricultural Engineering College and Research Institute, TNAU, Coimbatore, Tamil Nadu, [3] ICAR-Indian Institute of Soil and Water Conservation, Research Center, Nilgiris-643004, Tamil Nadu
[4] E-mail: dcsahoo@gmail.com

ABSTRACT

Hill slopes under tea plantation without any soil and water conservation measures in the initial stages is under the risk of land degradation and unsustainability due to excessive runoff and enormous soil erosion. Therefore, it is required to put into effect various conservation measures and evaluate the effect as related to the amount of runoff and soil erosion. In this context, a field study was conducted in the farmer's field in Nilgiris of South India, for evaluating the impact of different soil and water conservation measures under new tea plantation on sloping land of Nilgiris of South India. One year old B-6 tea clones were planted at double hedge spacing (135 cm x 75 cm x 75 cm) in two slopes (8-12 and 30-35%) with treatments *viz.*, contour staggered trenches (CST), vegetative barrier (VB), CST alternate with VB, CST with cover crop of beans and farmers' practice of plantation. Minimum runoff (14.6%) was observed from CST with cover crop of beans, followed by CST (15. 4%) under 8-12% slope range with exactly similar trend in runoff from the plots under 30-35% slope. Runoff retention in the trenches attributed to higher soil moisture in CST (25.6%) and CST with cover crop (24.8%) followed by CST with VB (22.8%) in 8-12 per cent slope with similar trend as well in 30-35 % slope during non-monsoon period. Tea canopy followed a similar trend to that of soil moisture content with maximum (39.5%) in thc CST with cover crop followed by CST (38.4%) and minimum in control just before the first harvest in 8-12% slope. Contrary to runoff, minimum soil loss was observed from CST (4.9 and 6.9 t ha^{-1} yr^{-1}) followed by CST with cover crop of beans (5.3 and 7.3 t ha^{-1} yr^{-1}) under 8-12% and 30-35 %, respectively. In adding together, either CST alone or in combination with cover crop is recommended for hill slope management through effective soil and water conservation under new tea plantation.

Keywords: Conservation measures, Contour staggered trenches, Cover crop, Double hedge, Vegetative barrier

Introduction

Soil erosion is a serious problem and major contributor to the soil loss in the new tea plantation areas. Hill slopes in the range of 10 to 50% and 500 to 2500 m above msl under tea plantation without any soil and water conservation measures in the initial stages is under high runoff and enormous soil erosion. In India, tea is grown over an area of 5780 km^2 mainly under rainfed sloping conditions (Madhu *et al.*, 2010) with annual rainfall varying from 1150 to 6000 mm. In Nilgiris of south India, tea is cultivated in areas having well-distributed annual rainfall of about 1200 mm. The problem of erosion in new tea plantations in Nilgiris is getting high over the years in the absence of any vegetative canopy and soil conservation measure. Therefore, soil loss assessment is a major concern in long run, as it affects the yield of green leaves and its sustainability.

Rainfall parameters, slope and cultural practices influence the runoff and soil erosion in new tea plantation. The farmers' practice of tea plantation, involves uprooting of bushes and shrubs for land preparation and planting the tea clones at desired spacing without following any conservation measures to control runoff and soil erosion. The disturbance of top soil in process of land preparation and plantation leads to considerable soil erosion through accumulated runoff along the slope length. Many tea estates in Darjeeling hills in the North-East India, the Nilgiris and the high range in south India have recurrently suffered from soil erosion problem. In Sri Lanka, Krishnarajah (1985) reported a loss of 20 tons of soil per hectare within six months in the absence of any earthworm. Soil erosion on sloping field in tea plantation area also leads to a lot of environmental problem. The on-site impacts caused to thin soil layer, deterioration in soil structure and decrease in soil nutrient (Zhang *et al.*, 2003). Evidence of soil degradation can be seen in the low soil organic matter content, cation exchange and water-holding capacity, highly acidic pH, high soil compaction, erosion, nutrient leaching, accumulation of xenobiotics and toxic aluminium present under intensive teal plantations (Senapati *et al.*, 1994a). Therefore, proper planning and implementation of soil and water conservation measures is very much required during plantation to reduce the surface runoff and soil loss. The importance of rainwater conservation, erosion control and maintaining soil productivity is increasing for sustaining agricultural production (Sahoo *et al.*, 2015). Vegetative barrier, cover crop, contour staggered trenches and combination of mechanical measures with any biological measures are some of the conservation measures adopted for controlling surface runoff and soil loss in the initial stages of tea plantation. The effects of various vegetative barriers on *in-situ* soil moisture conservation and in increasing crop yield were reported by several authors (Prasad *et al.*, 2005; Tiwari and Kurothe, 2006; Sahoo *et al.*, 2016b). Higher green tea yield with bio-engineering conservation measures under new tea plantation were recorded over control (Sahoo *et al.*, 2016a). Keeping these in mind, the study was conducted for hill slope management for controlling runoff and soil erosion during tea plantation in Nilgiris of South India.

Materials and Methods

Study Area

A field study was conducted during 2007-2009 in the farmer's field representing the most common area for tea in Nilgiris district (11° 26' 40" N and 76° 45' 58" E, 2150 m above mean sea level) of South India. The climate is temperate to sub-tropical with long term mean annual rainfall of 1276 mm with 79.4 % occurring during South-West and North-East monsoons. The mean monthly maximum and minimum temperatures are 22.1°C and 8.5°C occurring in April and January, respectively with mean annual temperature of 15.0° C. Mean relative humidity is about 76 % most of the time, favouring tea cultivation in the region.

Layout and Plantation

The experiment was laid out in a Randomized Block Design with four replications in 20 plots, each on two slopes (8-12 and 30-35%) to accommodate four farming methods and one farmers' practice of cultivation. Double hedge planting (135 cm x 75 cm x 75 cm) along the contour line accommodating 13,200 plants ha^{-1}was followed by using a triangular frame (0.75 m each side) for laying out the double hedge rows and plant spacing (Fig. 1). One year old B-6 tea clones were planted before the onset of North-East monsoon during the month of September, 2007 in humid climate and moist soil for better establishment of young tea plants.

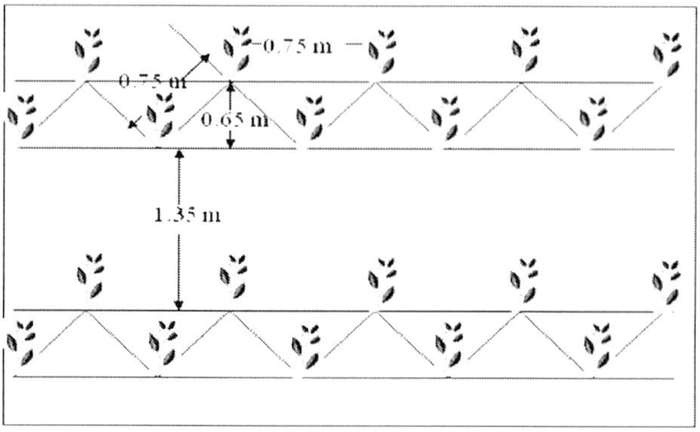

Fig. 1: Layout of tea plantation

Treatment Details

Treatments*	**Specification**
T_1: Contour staggered trenching(CST)	180 x 30 x 45 cm (length x width x depth)
T_2: Vegetative barrier (VB)	Two rows of geranium at P-P and R-R spacing of 30 cm
T_3: CST alternate with VB	180 x 30 x 45 cm alternate with VB at same spacing
T_4: CST with cover crop	180 x 30 x 45 cm with beans as a cover crop during monsoon period
T_5: Control	Farmers practice of cultivation without any conservation measures

*All the conservation measures are taken in the space between the double hedge rows

Runoff, Soil loss and Soil Moisture Monitoring

The multi-slot divisors are installed at the down end of the plots for runoff monitoring by measuring the runoff depth collected in each compartment in the stilling tank as well as in the collecting tank. The plots were separated to isolate the runoff from the adjacent plots using 60 cm height strips of metal sheet. The soil loss was estimated from runoff sample of 500 ml collected from stilling tank after stirring to determine the soil loss. Events wise runoff and soil loss analysis was made and expressed in mm and t ha^{-1} yr^{-1}, respectively. Soil moisture content was monitored by collecting soil samples using auger at 20, 40 cm and 60 cm soil depths during post monsoon period at monthly intervals and estimated gravimetrically to quantify the moisture content.

Results and Discussion

Runoff and Soil loss

The rainfall received during first year and second year are 986 mm and 1212 mm, respectively, thereby causing less total runoff in the first year than second year from all the treatments. However, the percentage of runoff to rainfall is higher in the first year in comparison to the following year from all the treatments in both the slopes showing the enhanced effect of conservation measures over time (Fig. 2). Average minimum runoff of 14.6% was obtained from CST with cover crop of beans, followed by CST (15.4%) under 8-12% slope range against maximum runoff (27.5 %) from control.

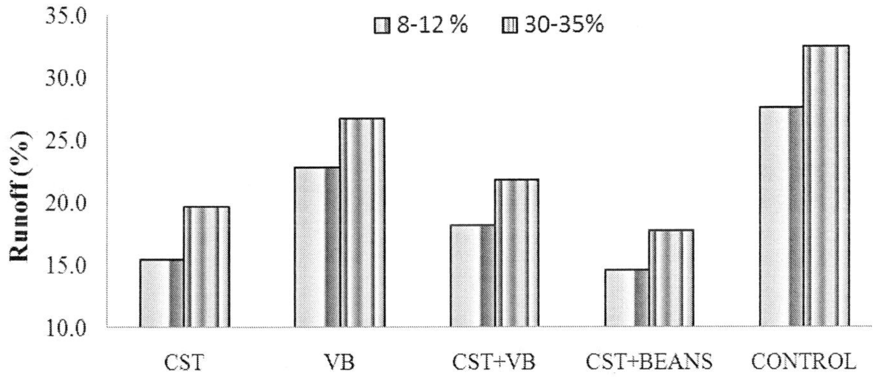

Fig. 2: Runoff under different conservation methods

Exactly similar trend in annual runoff values were obtained from the plots under slope range of 30-35% with minimum runoff from CST with cover crop of beans (19.4 % and 16.0% mm), followed by CST (20.9% and 18.4%) in contrast to maximum runoff of 33.6% and 31.4% from plots under farmers' practice in first and second year respectively. Therefore, it may be concluded that, CST is very much effective either alone or in combination in controlling runoff in the initial stages of tea plantation. Similar kinds of lower per cent of runoff to rainfall were reported under different conservation practices by Mane *et al.* (2009). Higher runoff volume and percent (17%) were found from all the treatments in 30-35 % slope in comparison to the respective treatments in 8-12% slope during both years. Low per cent of runoff to rainfall were also reported under different land use systems by Vinod *et al.* (2003).

Unlike runoff, minimum soil loss was observed from CST during both years with mean value of 4.9 t ha^{-1}yr^{-1}, followed by CST with cover crop of beans (5.3 t ha^{-1} yr^{-1}) under 8-12% slope (Fig. 3). This could be due to the disturbance of the top soil surface while taking the cover crop of beans than one time making of CST. However, by taking the cover crop of beans, followed by the inter-cultural operations, the soil became more porous due to root growth infiltrates more water in the soil profile and produced less runoff. Though maximum runoff was obtained from the plot under farmers' practice, the soil loss is maximum (10.0 t ha^{-1} yr^{-1}) from plot under vegetative barrier. Vegetative barrier alone could not reduce the erosion because of vulnerability of the top surface while taking the measure followed by inter-cultural operations. Therefore, the treatment of vegetative barrier is relatively less successful from the remaining treatments as far as the erosion is concerned.

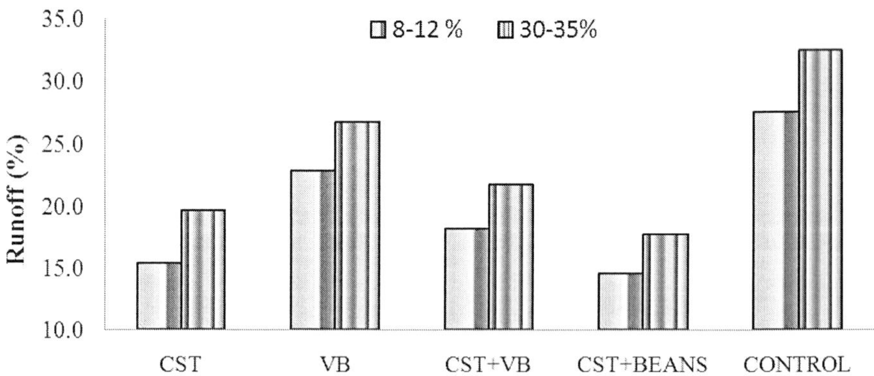

Fig. 3: Soil loss under different conservation methods

Similar trend in soil loss were observed from all treatments under 30-35% slope range. Minimum soil loss during the study was observed in the plot under CST (6.9 t ha^{-1} yr^{-1}), followed by CST with cover crop of beans (7.3 t ha^{-1} yr^{-1}) and CST alternate with vegetative barrier

(9.9 t $ha^{-1}yr^{-1}$) which shows that the CST either alone or in combination with the vegetative measures like cover crop of beans is the best in reducing erosion, when compared to other treatments.

Soil Moisture

Maximum mean soil moisture content of 25.1% was observed in CST followed by CST with cover crop of beans (23.3%) and minimum of 19.4% at 20 cm depth in VB under 8-12% slope during post-monsoon season. Similar result of soil moisture was observed with maximum (24.1%) in CST, followed by CST with cover crop to minimum of 18.6% in the plot of farmers' practice under 30-35% slope range. However, the close values of soil moisture contents between CST and CST with cover crop, followed by CST alternated with VB shows the effect of CST either alone or in combination in soil moisture conservation. Similar trend in soil moisture among all the treatments at other depths also. Higher soil moisture content can be attributed to runoff retention in the trenches, followed by absorption in the soil profile within the root zone. Moisture conservation and higher moisture availability in the soil profile have been reported by Singh *et al.* (2005) and Regar *et al.* (2009) due to conservation measures.

Canopy

Minimum variation in tea canopy among the treatments observed in the first year during monsoon and immediate after which may be due to the availability of enough soil moisture irrespective of any treatment effect in the slope range of 8-12 per cent. Afterwards, the variation in canopy due to treatment effect was realized. In March, the maximum canopy (34.5 per cent) was observed in the CST with cover crop of beans, followed by CST (32.7 per cent) with minimum in control and the same trend continued till June and upto first harvest. From the statistical analysis, the tea canopy from all treatments found significantly different from control at 5 per cent level of significance (CD (0.05) = 1.79). Unlike the observations made in the slope range of 8-12 percent, there is considerable variation in tea canopy between the control and remaining treatments from the beginning of the observation in slope range of 30-35 per cent. But within the extent of monsoon, variation is less among the treatments which widened afterwards. Maximum canopy was observed in CST (35.6 per cent), followed by CST with cover crop of beans may be due to the repeated disturbance while taking beans hampering the growth.

Conclusion

Hill slope management through soil and water conservation measures in new tea plantation are highly effective in controlling runoff and soil erosion. Implementation of CST with cover crop of beans and CST helps in maximum reduction in runoff due to temporary water storage in the trenches, followed by absorption, resulted in minimum erosion and better soil moisture in the root zone of the plants. Therefore, either CST alone or in combination with cover crop of beans are recommended for soil and water conservation under new tea plantation in the hill slopes.

References

Krishnarajah, P. 1985. Soil erosion control measures for tealand in Sri Lanka. *Sri Lanka J. of Tea Sci.,* 54(2): 91-100.

Madhu, M., Sahoo, D.C., Sharda, V.N. and Sikka, A.K. 2010. Rainwater-use efficiency of tea (*Camellia sinensis* (L.)) under different conservation measures in high hills of South India. *Applied Geography,* 31: 450-455.

Mane, M.S., Mahadkar, U.V., Ayare, B.L. and Thorat, T.N. 2009. Performance of mechanical soil conservation measures in cashew plantation grown on steep slopes of Konkan. *Ind. J. Soil Cons.,* 37(3): 181-184.

Prasad, S.N., Singh, R.K., Shakir Ali and Prandiyal, A.K. 2005. Comparative performance of grass barriers on erosion and crop yields in medium black soils of Kota. *Ind. J. Soil Cons.,* 33(1): 58–61.

Regar, P.L., Rao, S.S., & Joshi, N.L. (2009). Effect of *in-situ* moisture conservation practices on productivity of rainfed taramira in arid Rajasthan. *Ind. J. Soil Cons.,* 37(3): 197-200.

Sahoo, D.C., Madhu M., Bosu S.S. and Khola O.P.S. 2016a. Farming methods impact on soil and water conservation efficiency under tea [*Camellia sinensis* (L.)] plantation in Nilgiris of South India. *Intl. Soil and Water Cons. Research,* 195–198.

Sahoo, D.C., Madhu, M., Santhana Bosu, S. and Khola, O.P.S. 2016b. Influence of conservation practices on soil moisture and growth performance of new tea [*Camellia sinensis* (*L.*)] plantation in Nilgiris. *Ind. J. Soil Cons.,* 44(3): 297-301.

Sahoo, D.C., Madhu, M., Muralidharan, P. and Sikka, A.K. 2015. Land management practices for resource conservation under vegetable cultivation in Nilgiris hills ecosystem. *J. of Environ. Biology*, 36: 1039-1044.

Senapati, B.K., Panigrahi, P.K. and Lavelle, P. 1994a *Macrofaunal status and restoration strategy in degraded soil under intensive tea cultivation in India. In Transactions of the 15th World Congress of Soil Science*, ISSS, Acapulco, Mexico, 4A: 64-75.

Singh, R.K., Prasad, S.N., Ali, S., Kumar, A., Singh, K.D., Prasad, A. and Parandiyal, A.K. 2005. On-farm evaluation of conservation measures to performance of rainfed crops in semi-arid region. *Ind. J. Soil Cons.*, 33(2): 141-143.

Tiwari, S.P. and Kurothe, R.S. 2006. Effect of vegetative barriers on soil and nutrient losses at 2 per cent slope on agricultural lands of reclaimed Mahi ravines. *Ind. J. Soil Cons.*, 34(1): 37-41.

Vinod, V.R., Anwarulla, M. and Vishwanath, D.P. 2003. Runoff and soil loss under different land-use systems in the Western Ghats of Karnataka. *Ind. J. Soil Cons.*, 31(2): 131-138.

Zhang Y., Zhang H., Peng Bu-zhuo and Yang H. 2003. Soil erosion and its impact in Yixing tea plantation of Jiangsu Province. *Chinese Geographical Science*, 13(2): 142-148.

□□□

Chapter 18

Mechanization Issues and Options of Conservation Agriculture in NEH Region of India

Hijam Jiten Singh*, H. Dayananda Singh, Arvind Kumar and B.K. Sethy

Division ofAgril.Engg., ICAR-RC-NEH Region, Umiam-793103, Meghalaya, India, *Corresponding author E-mail: hijam_jiten@yahoo.co.in

ABSTRACT

Conservation agriculture (CA) is an increasingly adopted production system to meet the goals of sustainable crop production intensification in feeding a growing world population whilst conserving natural resources. Mechanization (especially no-till seeding and weed control) is a key input for CA. Equipment is available for the full range of power sources, from human and animal power to two and four-wheeled tractors. Manual and animal power equipment are invariably used on small farms and hilly areas, as they are usually small, light-weighted, simple in design and ease in maintenance. For weed management, all options should be explored including physical, biological and chemical control. Small and marginal farming is the mainstay of food production systems in NEH region and smallholder farmers often have difficulties in making the necessary investments. Farm equipment for the region, largely labour intensive and performed manually, must suit the hilly terrain and small farm sizes. Existing locally evolved tools give low output and involve excessive drudgery. Availability of CA equipment is a constraint. Therefore, local manufacturer should be encouraged and supported, whenever possible as machinery can be produced that is appropriate for local circumstances. A more immediate solution is to equip and train CA service provision entrepreneurs. With the right equipment, selected for the needs of their local farmers, and the right technical training so that such entrepreneurs can make a livelihood by supplying quality CA and other mechanization equipment.

Keywords: CA mechanization, small and marginal farmers, NEH region.

Introduction

Agriculture in north-east India, because of its diverse topography, altitude and climatic conditions, offers a huge scope for cultivation of wide variety of agricultural crops. Small and marginal farming is the mainstay of food production systems and the sustainable development of these systems is the key to ensuring long term food security in the region. Despite agriculture is the main economic activity in the region, modernization of agriculture has escaped this region as evidenced by poor adoption of modern technologies, low consumption of fertilizers and other indicators of growth. One of the most important reasons for this slow growth is the lack of location-specific and system-based technologies. Presently, the smallholding farms of the region have limited access to production inputs and so, achieve low levels of productivity. One key production input, mechanization, is mostly neglected in farm productivity improvement efforts, which may be either due to social, financial or technological constraints.Long-term conventional agriculture, which is plough and hoe-based agriculture, is intensely time and energy-consuming and often results in soil degradation, organic matter depletion and increased production cost. Sloppy farmlands suffering from moderate to severe soil erosion have accelerated land degradation, which has caused permanent, irreversible loss of land productivity.

Increase in food production via a process of sustainable intensification will, necessarily, require the implementation of more natural resource-friendly production methods, for example, reduced and no-till farming, and this will require a major diffusion of novel mechanization technology **(Sims and Kienzle, 2015).** Conservation agriculture (CA) has been proposed to reverse this degradation in an effort to move towards sustainable cropping systems. CA is a crop production system, based on minimum soil disturbance, surface crop residue retention and species diversification through crop rotations and associations **(Sims, 2011)**. Integrating livestock, poultry and fishery production with CA cropping is an appropriate option because of cultural preferences and market conditions in the region. Further, integrated farming system has been promoting by ICAR and state department to improve livelihood security.

To achieve the implementation of CA, there is a range of mechanization options for planting through soil surface residues and these include machines designed for manual, draught animal, two-wheel and four-wheel tractor power sources. However, farm equipment for the region must suit the terrain and small farm sizes. Largely, agricultural operations are labour intensive and performed manually. Existing locally evolved tools give low output and involve excessive drudgery (Singh *et al.*, 2014). This paper reflects the various mechanization issues and options classifying them into manual, animal power, two-wheel and four wheel tractors operations.

Mechanization of CA

The principal mechanization requirements for smallholder CA are for no-till seeding and weed control. Equipment is available for the full range of power sources, from human and animal power to two and four-wheeled tractors (2WT and 4WT). Manual and animal power equipment is usually small, light-weighted, simple in design and easily manufactured, utilized and maintained. This equipment is invariably used on small farms and hilly areas.

Manual and Animal Power Options

Li seeder is a typical manual seeder for no-till seeding of maize and soyabean (Fig. 1). The operating handle contains the seed and a shoulder bag carries the fertilizer. The total weight of Li seeder is 2.2 kg and the working efficiency is 0.2-0.3 ha/day/person (He *et al.*, 2014). Manually operated jab planters are suitable for very small holdings and are available with both seed and fertilizer (Johansena *et al.*, 2012). This machine enable to seed into mulch-covered no-tilled soil effectively. Jambo direct seeder, animal traction direct seeder and angled single disc seed drill are common animal power seeders. They were designed to manage residue whilst seeding and fertilizing in no-tilled soils, so they normally have discs to cut through the surface residue and furrow openers to place the fertilizer and seed. However, disc penetration in wet residue and hard soil can be a limitation.

Zero-till Metallic Tip Dibbler

Li-seeder Jab Planter

Fig. 1: Manual no-till or minimum-till seeder

Two-wheeled Tractor Options

No till planters for two-wheeled tractors are becoming increasingly demanded in Asian and African CA systems, such as ACIAR Rogro and strip-tillage machines (Esdaile, 2011). ACIAR-Rogro tined seed drill is a three-row tined planter with adjustable row spacing (Fig. 2). Attempt has been made on alternative soil engaging components, in addition to the standard chisel tine that has been fitted to date. These include double offset and single disc openers adjustable for offset and tilt. CA planters suitable for smaller farms include the Turbo Happy Seeder and chain-driven residue management system planters (Jat *et al.*, 2013). The machine has been designed specifically to deal with heavy rice residue (upto 10 ton/ha) and to direct seed wheat into this immediately after the rice harvest. Burning the straw is the other option and this produces both atmospheric pollution and loss of soil organic matter. The Turbo Happy Seeder has a set of straw management flails immediately in front of the chisel tine openers. These clear the straw and allow direct seeding to take place before the residue falls back to cover the soil. National zero-till multi-crop planter is attached behind a 2WT rotary tiller with the aid of clamps.

ACIAR-Rogro tined seed drill

Happy seeder coupled with 2WT(H.S. Sidhu)

Fig.2: Two-wheeled tractor for seeding

Tine and disc openers

There is also an option towards selection of tine and disc openers, such as Stubble Star double disc opener, 'Daybreak' single disc opener technology, etc. to reduce soil disturbance during the seeding process (Fig. 3). The discs of stubble star double notched disc opener are angled at 7° to each other and have an undercut angle of 3.5° to improve soil conditions in the base of the furrow (Desbiolles, 2011).

Stubble star double disc opener

Daybreak' single disc opener

Fig. 3: Tine and disc opener for minimum soil disturbances

For weed management, all options should be explored including physical, biological and chemical control. Manually powered mechanical control options include shallow scraping with sharp hand-hoes, hand pulling and slashing which are suitable for very small areas. As holdings become larger, then animal traction and tractor power can be used for knife rolling. A precision farming technology "The WeedSeeker" (Fig. 4) technology is attractive because of the reduced applications of herbicides, but it also allows applications of specific, and usually more expensive, herbicides to problem weeds showing signs of resistance to conventional herbicides, such as glyphosate. The sensors of the technology work on the principle of reflecting red and near infra red light (from a light-emitting diode - LED) off a weed 600 mm beneath. When a green plant is identified (through an on-board analysis of the reflected light), the spray nozzle is activated above the offending weed (Sims, 2011).

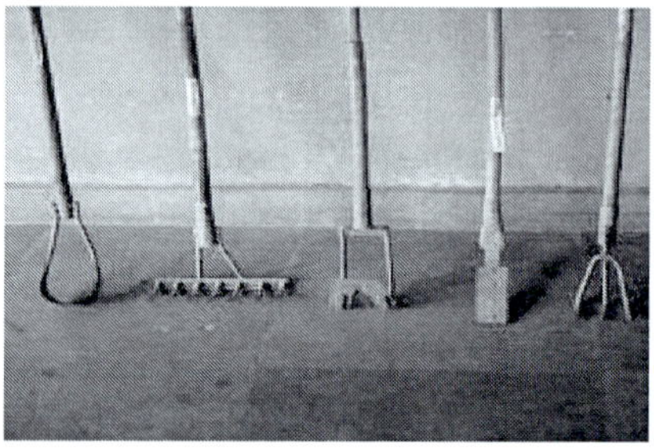

Manual long handle weeders

WeedSeeker

Fig. 4: Manual hand tools and WeedSeeker Technology

Mechanization Issues of CA

Improving smallholders' access to crucial mechanization inputs is frequently fraught with difficulties as the adoption of any innovation must be seen as a useful and profitable investment from the perspective of the farmer. This pre-requisite would then trigger the necessary demand for these innovations which should lead to an increased or stabilized supply. However, very often this demand is non-existent due to the low income levels of small farmers which in turn also lead to only very rudimentarily developed mechanization input supply chains in rural areas. This situation is referred to as 'the vicious cycle of mechanization development'. It will require broad action among all actors involved in rural development and sustainable intensification programmes from the farmers' level upto and including policies to disintegrate the cycle (Kienzle and Sims, 2014).

Mechanization usually involves a lot of investment. So, a smallholder with, typically, under two hectares of land, will be reluctant to be the sole investor in a machine which has the potential to operate over a much larger area. As the farmers of the region are mostly tribal and economically backward, costly machines are not affordable and will militate to adopt such machinery. The poor smallholder families in the region, who are struggling to emerge from subsistence farming, will usually have difficulty in amassing the resources necessary for machinery purchase.

In NEH region, local availability of CA equipment (especially planters) is extremely limited. Supply is frequently initiated by externally funded aid programmes and projects and local supply chains are generally not adequately developed in this part of the country. This can lead to the situation, where farmers outside the programme target group are unable to access the machinery, even if they are convinced that it would be beneficial, as there is no local supply network. Therefore, local manufacture should be encouraged, whenever possible, as machinery can be produced that is appropriate for local circumstances. The most attractive option to improve access by smallholders to CA mechanization would be to offer the service from well-equipped and well-trained local service providers. Further technical training should be conducted, such as hands-on and predominantly practical in the field e.g. calibration of seeders and sprayers, field operation of rippers, seeders and sprayers, routine maintenance, Servicing, e.g., of 2WT engines and transmissions.

Conclusions

Mechanization is a major input to CA and well-functioning equipment must be on the market for wider adoption. For small and medium scale farmers, there are advances in manual, draught animal powered and tractor powered options (especially with 2WTs). Although a variety of no-till equipment has been developed and fabricated, some improvements can be made to suit a range of crops and soils. To adopt and popularise hilly terrain, small farm size, and economic constraints of NEH region, CA machinery need to be light weighted, simple, affordable, and suited to low power source. Local manufacture should be encouraged, whenever possible, as machinery can be produced that is appropriate for local circumstances. Further, service provision from well-equipped & well-trained local service providers should be met available and technical training, such as hands-on and predominantly practical in the field e.g. calibration of seeders and sprayers, field operation of rippers, seeders and sprayers, routine maintenance, Servicing, e.g., of 2WT engines and transmissions should also be conducted.

References

1. Sims, B.G. (2011). Some advances in mechanization options for Conservation Agriculture systems: reflections from the V World Congress of Conservation Agriculture, Brisbane, Australia,

September 2011. Available at: http://iagrehost.org/sites/iagre.org/ files/ landwardsextra/Reflections%20paperv2. pdf. Assessed on 18/ 09/2015.

2. Desbiolles, J.M.A. (2011) Disc seeders in conservation agriculture: an Australian survey. 5th World Congress on CA. pp18-19. Available at: http://aciar.gov.au/WCCApapers. Accessed on 16/01/2017.

3. Esdaile, R.J. (2011) Conservation farming implements for two wheel tractors. 5th World Congress on CA. pp.364-365. Available at: http:/ /aciar.gov.au/WCC Apapers. Accessed on 16/01/2017.

4. He, J., Zhang Z.Q., Li, H.W. and Wang, Q.J. (2014) Development of small/med ium size no-till and minimum-till seeders in Asia: A review. Int J of Agric. & Biol. Engg. 7(4):1-12.

5. Jat, M.L., Kapil, Kambo,j B.R., Sidhu, H.S., Singh, M., Bana, A., Bishnoi, D., Gathala, M., Saharawat, Y.S. and Kumar, V. (2013). Operational Manual for Turbo. Happy Seeder-Technology for Managing Crop Residues with Environmental Stewardship; International Maize and Wheat Improvement Center (CIMMYT), Indian Council of Agricultural Research (ICAR): New Delhi, India, pp. 28.

6. Johansena, C., Haque, M.E., Bell, R.W., Thierfelder, C. and Esdile, R.J. (2012) Conservation agriculture for small holder rainfed farming: Opportunities and constraints of new mechanized seeding system. *Field Crop Research*, 132: 18-32

7. Kienzle, J. and Sims, B.G. (2014) Agricultural mechanization strategies for sustainable production intensification: Concepts and cases from (and for) sub-Saharan Africa. Available at: http://www. clubofbologna.org/ew/documents/31b KNRKienzlemf.pdf. Accessed on 16/01/2017.

8. Sims, B.G. and Kienzle, J.(2015) Mechanization of Conservation Agriculture for Smallholders:Issues and Options for Sustainable Intensification. *Environments,* 2: 139-166.

9. Singh, H.J., De, D., Sahoo, P.K., and Iquebal, M.A. (2014) Development and Evaluation of Self- propelled Multicrop Planter for Hill Agriculture. *Journal of Agricultural Engineering*, 51(2): 1-8.

❑❑❑

Chapter 19

Phenological Growth Stages of Guava Cultivars under Subtropical Conditions

Karma Beer*, Vikas Yadav[1], S. Rajan[2]

* ICAR- Indian Institute of Soil and Water Conservation, Koraput, Odisha

[1] ICAR- Central Institute of Arid Horticulture, Bikaner, Rajasthan

[2] ICAR- Central Institute of Subtropical Horticulture, Lucknow

Introduction

Guava fruit (*Psidium guajava* L.) or poor man's apple is an evergreen fruit native to tropical Central America, considered a hardy fruit crop besides being rich in nutrients (Morton, 1987). It has a tendency to produce two to three crops annually offering sustained livelihood options throughout the year. Uttar Pradesh is an important guava growing state in India and best quality guavas are grown here. The best temperature for guava cultivation ranges from 15 to 30 ^{0}C, with an annual average temperature of 18.8 ^{0}C (Somarriba, 1985b). During vegetative growth-guava tree undergoes several physiological changes that are identified because of external signs, such as shoot growth, bud swelling and break, increments in trunk diameter as well as reproductive phases like flower initiation, fruit setting and ripening. Nutritive value of guava is high. Therefore, it is an ideal fruit for nutritional security (Chadha, 2015).

The sequential study of all periodical events involved in a plant life is called 'phenology' (Volpe, 1992; Villalpando and Ruiz, 1993; Schwartz, 1999). Beginning and end of every developmental phase and stage contributes to evaluate plant growth rates (Torres, 1995). Phenological growth is the interval between two different phases (Villalpando and Ruiz, 1993). Different type of phenological indicators are used to moniter and evaluate plant development. Blooming time and fruit ripening is the most significant stage in plants. The interval between blooming and fruit set is one of the most important phenological stages (Villalpando and Ruiz, 1993).

In herbaceous and woody plants a new nomenclature was described by Bleiholder *et al.* (1989). The BBCH general scale applicable was specific to different botanical families into a general scale applicable to all plants. A detailed version of BBCH scale was published by Stauss (1994). First stages refer to its major stage using value from zero to nine and second digit also scaled from zero to nine that relates to its secondary stage of plant growth.

The several authors classify different plant families; like cereals by Zadoks *et al.* (1974), rape, bean and sunflower by Lancashire *et al.* (1991), beet by Meier *et al.* (1993), potato by Hack *et al.* (1993), stone and pip fruits, redcurrant and strawberry by Meier *et al.* (1994), Japanese loquat tree (*Eribotrya japonica* Lindl.) by Martinez-Calvo *et al.* (1999) , vine, several vegetables and citrus species by Lorenz *et al.* (1994), Feller *et al.* (1995) and Agusti *et al.* (1995), respectively.

The objective of this study was to study the phenological stages of different guava cultivars in sub-tropical condition. The techniques of recording the phenological stages were allowing the exact stage of evolution to be determined at any time during these physiological changes.

Materials and Methods

The experiment was conducted at ICAR- Central Institute for Sub-tropical Horticulture Rehmankhera, Lucknow in randomized block design with three replications in order to analyze phenological stages of guava. The experiment was done with six treatments comprising of six commercial guava cultivars ('Lalit', 'Shweta', 'Allahabad Safeda', 'Chittidar', 'Dharwar' and 'HPSI-46'). Phenological stages were determined for each cultivar. Three branches, which had almost similar morphological conditions, were selected for recording phenological stages regularly. Age of different guava cultivars was from 10 to 12 years. Each tree was planted at 5×5 meter spacing; soil of orchard was sandy loam. Guava plants were irrigated by ridge and furrow method. Highest and lowest mean monthly temperature was recorded 39.8 and 5.3 °C, respectively. Relative humidity ranged from 63.3 to 88.3 per cent and total annual rainfall of 553.2 mm was recorded (Anonymous, 2015). Five plants of each guava cultivars were selected and three branches were marked as replication for data collection; hence thirty guava plants with 90 branches were selected for the study.

Gil-Albert (1989) reported that techniques of recording the phenological stages permitted the study of growth and development of vegetative and reproductive organs of different plant species, allowing the exact stage of evolution to be determined at any time during these physiological changes. Frequency of visit in guava orchard ranged from one week to two weeks, depending upon the change in phenology of guava. Good quality photographs were taken for later analysis. There were rapid growth in guava from bud break to fruit enlargement; hence the orchard was visited one to two days interval. Fruit diameter was measured by vernier calliper. Phenological stages of guava of different cultivars were observed according to BBCH (Biologische Bundesanstalt, Bundessortenamt und Chemische Industrie) scale and different stages data were recorded in days.

Results and Discussion

Phenological studies are very useful to know the periodicity of life-cycle of fruit trees for better agronomical management. The extended BBCH (Biologische Bundesanstalt, Bundessortenamt und Chemische Industrie) scale can provide more accurate and detailed description of the most important phenological events than any of the other coding systems for guava (Table 1). Phenological stages (in days) with their respective codes are given in Table 1 and Fig. 1. The development stage in different guava cultivars stats from swelling of bud to ripening of fruit. Collected data was subjected to analyze in excel sheet. Details of phonological stage and its description are given in Table 2.

Table 2: Details of phonological stage

BBCH scale	Phonological stage	Description
00	Winter bud	The bud is greenish brown and completely closed. It is closely linked to the twig
01	Bud swelling	The bud swells and becomes greenish
02	Bud growth begins	The bud gets longer, and scales start opening
11	First leaves sprouting	The first leaf appears and becomes visible
15	More leaves unfolded	Not yet at full size
19	Leaves completely developed	Leaf growth completed
51	Appearance of flower buds	Calyx becomes visible. Internodes lengthening stops
55	Flower buds visible	Sepals still closed
57	Flower petals elongating	Sepals slightly open petals just visible
59	Sepals totally open	Sepals fully extended so that petals can open
65	Full flowering	At least 50% of flowers open. First petal falling
67	Petal fall	Flowers fading and most petals collapse
71	Fruit setting	Fruit size upto 10 mm
78	Fruit growth	Fruit increase upto 80% of final size
81	Fruits colour changing	Beginning of ripening; fruit reaching its final volume. Colour changing from green to yellowish pale green
89	Fruit ripening	Fruit reached final volume and become completely yellow. Releasing pleasant aroma, and fruit right for consumption

In guava, vegetative growth normally includes bud, leaf and shoots development. Results from Table 2 revealed that from bud to first leaf development it takes about (10-12 days). It takes 9-12 days to complete leaf development. Appearance of flower buds takes 1-3 days and flowers bud became visible from 2-3 days. Elongation of flower petals takes 15-19 days in different cultivars. It was observed that sepals totally opened 8-10 days. Flowers 50% opened ranged from 2-3 days. Fruit set started from 2-3 days. Fruit growth was completed in 95-112 days in different cultivars (Table 1). Fruit colour changes from 5-7 days and fruit ripens in 6-10 days. Growth of guava fruit shows the double sigmoid pattern described by Araujo *et al.* (1997), Mercado-Silva *et al.* (1998) and Selvaraj *et al.* (1999). It is shown in Fig. 2, which have similar growth trend. The stages from fruit set to guava physiological maturity takes 119 to 130 days reported by Laguado *et al.* (1999) and Mercado-Silva *et al.* (1998). Somarriba (1985a) found in his study that guava reaches in maturity 90-150 days after anthesis.

Fruit growth of guava was monitorised from fruit set to fruit maturity. It is clear from the Fig. 2 that guava fruit follows the double sigmoid growth curve. During the entire study, total 130 days cycle and three stages were identified. First stage went to 55 days and it is characterized by an exponential growth due to cell division and enlargement of guava fruit (Garces, 1987) at the end of this stage guava reached 30 mm average fruit diameter. Second stage took from 56 to 80 days after fruit set. Guava growth clearly slowed down (up to 36 mm) because of cell division and enlargement reductions caused by endocarp hardening (Barcelo *et al.*, 1992). Physiological maturity of guava came 130 days after fruit set. Maximum growth of fruit diameter was recorded in this stage due to water and solute accumulations within cell vacuoles (Coombe 1976; Garces 1987). At this stage guava reached 57.5 mm diameter.

After identification of the phenological stages of guava, the different activities like pruning time, pesticide application, application of growth regulators, irrigation and maturity dates can be performed at appropriate time. Hence, knowledge of phenological stages of six guava cultivar can be useful for farmers.

Acknowledgements

All The authors are highly thankful to Director, ICAR- Central Institute for Sub-tropical Horticulture, Lucknow for providing the necessary facilities to carry out this research work.

References

Agusti, M., Zaragoza, S., Bleiholder, H., Buhr, L., Hack, H., Klose, R., Stauss, R. 1995. Escala BBCH para la descripcion de los estados fenologicos del desarrollo de los agrios (Gen Citrus). Levante Agricola 332: 189-199.

Anonymous, 2015. ICAR- Central Institute for Sub-tropical Horticulture-Lucknow, Annual Report-2015-16, p. 165.

Araujo, F., Quintero, S., Salas, J., Villalobos, J., Casanova, A. 1997. Crecimiento y acumulacion de nutrientes del fruto de guayaba (*Psidium guajava* L.) del tipo Criolla Roja en la planicie de Maracaibo. Rev. Fav. Agron. (LUZ) 14: 315–328.

Barcelo, J., Nicolas Rodrigo, G., Sabater Garcýa, B., Sanchez Tames, R. 1992. Fisiologýa Vegetal Ciencia y Tecnica. Piramide, Madrid, pp. 412–584.

Bescheribung nach der erweiterten BBCH-Skala, mit abbildungen. Nachrichtenblatt Deutsche Pflanzenschutzd 46: 141-153.

Bleiholder, H., van den Boom, T., Langeluddeke, P., Stauss, R. 1989. Einheitliche Codierung der phanologischen Stadien bei Kultur-und Schadpflanzen. Gesunde Pflanzen 41: 381-384.

Chadha, K.L. 2015. Guava. Handbook of horticulture, pp. 189-194.

Coombe, B. 1976. The Development of Fleshy Fruits. Ann. Rev. Plant Physiol. 27: 507–528.

Feller, C., Bleiholder, H., Buhr, L., Hack, H., Mess, M., Klose, R., Meier, U., Stauss, R., van den Boom, T., Weber, E. 1995. Phanologischen Entwicklungsstadien von Gemuse. l.Zwiebel- Wurzel-, knollenund Blattgemu¨se. Nachrichtenblatt Deutsche Pflanzenschutzd 47: 193-206.

Garces, G.E. 1987. Estudio anatomico y de los procesos de crecimiento del fruto del guayabo (*Psidium guajava* L.) Agron. Colomb. 4: 23–30.

Gil-Albert, F. 1989. Tratado de arboricultura frutal, Aspectos de la Morfologia Fisiologia del arbol Frutal, 3rd edition, vol. 1. Ediciones Mundi-Prensa, Madrid, 104 pp.

Hack, H., Gall, H., Klemke, T., Klose, R., Meier, U., Stauss, R., Witzenberger, A. 1993. The BBCH Scale for phenological growth stages potato (*Solanum tuberosum* L.) *In*: Proceedings of the 12th Dreijahrestagung European Geesellschaft, Kartoff, Paris, pp. 153-154.

Laguado, N., Marin, M., Arenas de Moreno, L., Araujo, F., Castro de Rincon, C., Rincon, A. 2002. Crecimiento del fruto de guayaba (*Psidium guajava* L.) del tipo Criolla Roja. Rev. Fav. Agron. (LUZ) 19: 273-283.

Lancashire, P.D., Bleiholder, H., van den Boom, T., Langeluddeke, P., Stauss, R., Weber, E., Witzenberger, A. 1991. A uniform decimal code for growth stages of crops and weeds. Ann. Appl. Biol. 119: 561-601.

Lorenz, D.H., Eichorn, H., Bleiholder, H., Klose, R., Meier, U.,Weber, E. 1994. Phanologische Entwicklungsstadien der Weinrebe (*Vitis vinifera* L. ssp. vinifera) Codierung und Bescheribung nach der erweiterten BBCH-Skala. Viticulture Enol. Sci. 49: 66-70.

Martinez-Calvo, J., Badenes, M.L., Llacer, G., Bleiholder, H., Hack, H., Meier, U. 1999. Phenological growth stages of the loquat tree. Ann. Appl. Biol. 134, 353-357.

Meier, U., Bachmann, L., Buhtz, E., Hack, H., Klose, R., Marlander, B.,Weber, E. 1993. Phanologische Entwicklungsstadien der Beta-Ruben (*Beta vulgaris* L.). Codierung und Bescheribung nach der erweiterten BBCH-Skala mit Abbildungen. Nachrichtenblatt Deutsche Pflanzenschutzd 45: 37-41.

Meier, U., Garf, H., Hack, H., Hess, M., Kennel, W., Klose, R., Mappes, D., Seipp, D., Stauss, R., Streif, J., van den Boom, T. 1994. Phanologische Entwicklungsstadien der kernobstes (*Malus domestica* Borkh. und *Pyrus communis* L.), des steinobstes (Prunus-Arten), der johannisbeere (Ribes-Arten) und der erdbeere (*Fragaria* x *ananassa* Duch.). Codierung und

Mercado-Silva, E., Benito-Bautista, P., Garcýa-Velasco, M. 1998. Fruit development harvest index and ripening changes of guavas produced in Central Mexico. Postharv. Biol. Tech. 13: 143-150.

Morton, J.F. 1987. Fruits of warm climates. Creative Res. Syst. Inc. 91-102.

Salazar, D.M., Melgarejo, P., Martinez, R., Martinez, J.J., Hernandez, F. and Burguera, M. 2006. Phenological stages of the guava tree (*Psidium guajava* L.). *Sci. Hortic.*, 108: 157-161.

Schwartz, M.D. 1999. Advancing to full bloom: planning phonological research for the 21st century. Int. J. Biometeorol. 42(3): 113-118.

Selvaraj, Y., Pal, D., Raja, M., Rawal, R. 1999. Changes on chemical composition of guava fruits during growth and development. Indian J. Hort. 56(1): 10-18.

of 790 kg ha^{-1} under moisture stress conditions in acid soils of Ri-Bhoi district of Meghalaya. The present success story in the farmers' field indicates that rapeseed is a climate resilient crop which can be grown in the residual soil moisture and the farmers of NEH region in general and Meghalaya in particular, could get nutritional as well as livelihood security through increased oilseed productivity and reduced cost of cultivation with less effort.

Keywords: Oilseeds, rapeseed, North-Eastern hill region, zero tillage, and livelihood security, etc.

Introduction

Oilseed is the second most important crop group which determines agricultural economy, next to cereals. India is world's fourth largest edible oil economy after the U.S., China and Brazil. Globally, it contributes almost 6 per cent of global vegetable oil production, 14 per cent of vegetable oil imports and 10 per cent of edible oils (Mir *et al.*, 2010). Rapeseed and mustard contributes 32 per cent of the total oilseed production in India and it is the second largest oilseed group crop after groundnut. It contributes to the 4.1 M ha area and 6.6 Mt production of total oilseed and mustard of India with average productivity of 1182 kg ha^{-1}.

The North-Eastern Hill (NEH) region of India is primarily under the acidic soil zone with high rain fall area. Majority of the fields in the region are situated across the hilly slopes with shallow and gravely soils. Shifting cultivation (*Jhum*) is one of the pre-dominant agriculture practised by the tribal inhabitants. Utilization and improvement of soil resources is of paramount importance for enhancing crop production.Oil seed crops are important for NEH region, as oil is utilized throughout the region for cooking and frying purposes. The oil seeds are extensively used as condiments in preparation of pickles, curries and vegetables. The leaves of young plants are also used as green vegetables as it supply enough sulphur (S) and minerals in the diet. However, the per capita availability of oilseed in the region is 8 g day^{-1} as against moderate requirement of 50 g day^{-1}.The population growth in this region is rapid and food demands are speedily increasing but large chunk of the area is under fallow after the rainy season crop. This pressurizing the growers to produce more food per unit area, in phosphorus (P) deficient soils (Choudhary *et al.*, 2013).

In acid soils, P adsorption is generally attributed to hydrous oxides of iron and aluminium. There is great possibility that some natural phosphates of aluminium or iron (such as variscite and strengite) may formed in these soils. Therefore, P is the most limiting nutrient for crop production in acid soil. Oilseed cultivation in the NEH region faces other several constrains, such as water scarcity during post-monsoon season, lack of irrigation facilities, short time lag after rice harvest for seed sowing and lack of soil test based fertilizer recommendations. Rice-fallow cropping system can be substituted by introducing short duration oilseed crops like toria (*Brassica campestris* var. toria). Toria not only provides the additional yield, better land use efficiency but also help to improve the livelihood and nutritional security. North-East India comprises more than 81 per cent acidic soils out of 26.22 M ha land (Sharma *et al.*, 2006).

The production of oilseed under acidic soils of North-East India can be augmented by adopting the modern crop production techniques and sustainable use of available natural resources, use of chemical fertilizers to supply essential plant nutrients in required amount and in available form which increase the uptake of nutrient under acidic soils of hilly regions. Use of mulching techniques because under hilly conditions, soil water is the major constraints and also the weeds are the universal problem in agriculture, but under upland condition and high rainfall areas creates more weed problem. Use of green manuring techniques and also liming practices will helps in reduction of iron toxicity to oilseed crops under acidic condition.

The effects of mulching on soil moisture depend on precipitation and climatic factors. Mulching favourably influences soil moisture regime by controlling surface evaporation rate in summer and winter, mulching conserves soil moisture by reducing the evaporation rate. Mulches improve soil-moisture retention capacity as well as soil structure and suppress weed growth (Mutetwa and Mtaita, 2014). The soil-moisture variation is not, however, same under different mulching materials. The amount of soil-moisture conservation under different mulching materials differs in different soil types and climatic conditions. In general, the mulching treatments store higher soil moisture compared to the bare soil (no mulch) (Zhao *et al.*, 2012).

Application of organic mulches decreases the soil temperature, as it creates barrier for penetration of solar radiation to soil surface. Paddy straw is easily available in North-East India during November-

December and it is an eco-friendly source of organic mulch. It is the cheap and benign technology for rural farmers. It also conserves soil moisture, maintaining favourable soil moisture regimes by reducing evaporation loss from soil. Rice is a core cash crop and staple food in many Asian countries (Rao *et al.*, 2007; Kumar and Ladha, 2011). The biological and chemical properties of soils play an essential role for the regulation of organic matter decomposition, carbon sequestration and nutrient mineralization that are crucial for soil health which leads to increased oilseed production.

Measures for Augmenting Oilseed Production in NEH Region

i) Adoption of zero tillage practice

Water scarcity during post-monsoon seasons and lack of irrigation facilities, short time lag after rice harvest for seed sowing and high incidence of pests and diseases in late sown crops are the major constraints for oilseed cultivation in the NEH region. As a result, only mono-cropping of rice is practiced and the farmers left their land fallow. The Central Agricultural University, Imphal in collaboration with Directorate of Rapeseed-Mustard Research, Bharatpur implemented an extension project entitled, "Augmenting Rapeseed-Mustard Production of Tribal Farmers of North-Eastern States for sustainable livelihood security" under the Tribal Sub-Plan (TSP) during *rabi*, 2011. Yield performance of rapeseed, varieties M-27 and TS-36, yellow sarson, Ragini and mustard varieties, Pusa Agrani, Pusa Mahak, NRCHB-101 and NPJ-112 were evaluated in 55 ha under Zero tillage cultivation and compared with crops grown in 40 ha under conventional tillage (Plate 1 & 2).

Besides, no tillage (zero tillage) practice, use of 4 (four) bee colonies ha^{-1} during crop bloom for pollination and spray of botanical pesticides without affecting pollinators population and production of organic honey, were demonstrated. Since there was no rain throughout the crop period, the growth and yield parameters in all the rapeseed-mustard varieties were better in zero tillage than conventional tillage due to residual soil moisture after rice harvest. Among the rapeseed varieties, yellow sarson, Ragini gave the maximum average yield of 1000 kg ha^{-1} (range: 800 to 1400 kg ha^{-1}), whereas, NRCHB-101 among mustard varieties gave maximum average yield of 1020 kg ha^{-1} (range: 800 to 1100 kg ha^{-1}) both in zero tillage cultivation. In all, 172 farmers across 9 (nine) villages of Imphal East District involved in the project improved their income by getting average net profit of Rs.27,388 ha^{-1} including cost of honey within three and half months with a low investment of Rs.13,412 ha^{-1}. By observing the standing crop in thefield altogether 1419 farmers across 50 villages in10 (ten) districts of 3(three) North Eastern States namely Manipur, Mizoram and Arunachal Pradesh adopted this technology and the area coverage under Zero tillage cultivation of rapeseed-mustard increased to 1010 ha during *rabi*, 2012 and 2013.

Under the water stress situation where there was no rainfall during the crop period of rabi, 2012, M-27 among rapeseed varieties and YSH-401 among yellow sarson varieties and NRCHB-101 among mustard varieties gave maximum average yield of 600, 1000 and 1190 kg ha^{-1}, respectively under zero tillage cultivation. Similarly, during *rabi*, 2013 under the water stress situation, TS-38 among rapeseed varieties, YSH-401 among yellow sarson varieties and NRCHB-101 among mustard varieties gave average yield of 790, 950 and 1180 kg ha^{-1}, respectively under zero tillage cultivation. It indicates that rapeseed-mustard is a climate resilient crop which can be grown without water in the residual soil moisture. By adopting zero tillage, the farmers could increase the productivity, reduced cost of cultivation thereby increasing the cropping intensity and earning an additional income for themselves with less efforts (Plate 1). Zero tillage also helps in timely sowing (October-November), conserves soil moisture and requires less water, saves tillage cost and time, and the soil is protected from erosion due to the retention of surface residues and reduce organic matter depletion.

The improved version of this zero tillage cultivation with bee pollination and no chemical method of plant protection may be recommended to the resource poor farmers of the North-Eastern Region in the context of climate change.

Plate 1 & 2: Zero tillage cultivation of Rapeseed, TS-38

Plate 3: Training Programme on Zero tillage cultivation of Rapeseed

ii) Managing phosphorus availability in acid soils

For quantifying right P doses in acidic soils to get optimum production of oilseed crops, Maurya and Swami (2017) determined critical limits of available phosphorus for Rapeseed (*Brassica campestris* var. *toria*) in acidic soils of North-East India. The investigation was carried out during *rabi* season of 2016-17 using 6 levels of P (0, 30, 60, 90, 120 and 150 mg kg^{-1} soil) applied through two sources *viz.* single super phosphate (SSP) and rock phosphate (RP) in different acidic soils *viz.* Alfisol and Inceptisol. They observed that the highest mean dry matter yield of rapeseed (16.11 g pot^{-1}) was recorded with the application of 120 mg of P kg^{-1} of soil (Table 1). The further increase in the dose of P by 150 mg of P kg^{-1} soil significantly decreased the dry matter yield by 4.59 per cent over 120 mg of P kg^{-1}. The dry matter yield produced with RP was of lower order as compared to SSP at each levels of P irrespective of soil type. The Inceptisol has produced significant higher amount of dry matter as compared to Alfisol irrespective of P sources. The highest dry matter yield (17.81 g pot^{-1}) was observed in Inceptisol with 120 mg kg^{-1} soil under SSP, whereas the lowest dry matter yield (5.57 g pot^{-1}) was recorded in Alfisol with control under RP. Similarly, significant difference in dry matter yield of rapeseed was found between the two sources of P irrespective of the soil type. In Alfisol, the SSP has produced 7.71 per cent higher dry matter yield over RP in control, whereas, in Inceptisol SSP has produced 2.56 per cent higher dry matter yield in control pots.

Table 1: Effect of different levels and sources of P on rapeseed dry matter production in acidic soils of North-East India

P levels mg kg^{-1} soil	Alfisol		Inceptisol		Mean
	RP	SSP	RP	SSP	
0	5.57	6.00	7.02	7.20	6.45
30	9.34	10.30	10.77	11.89	10.58
60	12.12	13.20	14.14	15.61	13.77
90	14.21	15.00	16.13	17.68	15.76
120	14.44	15.20	17.02	17.81	16.11
150	14.12	14.90	15.80	16.65	15.37
RP	12.56	Alfisol	12.03	S.E(m)±	0.14
SSP	13.45	Inceptisol	13.98	CD (p=0.05)	0.39
For Source of P and Soils		Interaction			
		S x So	S x L	So x P	S x So x L
S.E(m)±	0.08	0.11	0.19	0.19	0.28
CD (p=0.05)	0.23	0.32	0.55	0.55	0.78

Further, they also found that the application of P with different levels significantly improved the P uptake by rapeseed plants. The overall mean P uptake in the control was recorded to be 9.83 mg P pot^{-1} in control, which increased significantly by 4.9 times with the application of 120 mg P kg^{-1} soil (Table 2). The P uptake by rapeseed plant irrespective to two different soils showed an increasing trend from no P application to the 120 mg P kg^{-1}of soil and then reduced. The highest P uptake was recorded in Inceptisol through SSP as a source i.e. 59.37 mg pot^{-1} compared to RP (49.52 mg pot^{-1}) whereas in case of Alfisol, highest value of P uptake was recorded through SSP (45.23 mg pot^{-1}) compared to RP (39.06 mg pot^{-1}).

Table 2: Effect of phosphorus sources and levels on P concentration (per cent) rapeseed in acidic soils of North-East India

P levels	Alfisol		Inceptisol		Mean
	RP	SSP	RP	SSP	
0	0.12	0.14	0.16	0.18	0.15
30	0.14	0.17	0.21	0.23	0.19
60	0.18	0.20	0.25	0.27	0.23
90	0.24	0.25	0.28	0.31	0.27
120	0.27	0.29	0.30	0.33	0.30
150	0.28	0.30	0.31	0.35	0.31
RP	0.23	Alfisol	0.22	S.E(m)±	0.01
SSP	0.25	Inceptisol	0.26	CD (p=0.05)	0.02
For Source of P and Soils		Interaction			
		S x So	S x L	So x P	S x So x L
S.E(m)±	0.001	0.01	0.01	0.01	0.01
CD (p=0.05)	0.01	0.02	0.03	0.03	0.04

iii) Lime application

Use of lime is often recommended to increase the crop productivity in acidic soils. To ascertain the individual and synergistic effects of lime, NPK and farmyard manure (FYM) application on ground nut productivity, a field experiment was undertaken on acid soils of West Tripura. Application of recommended dose of NPK (20:60:42 kg ha^{-1} of NPK) along with lime (10 per cent of actual LR was followed) resulted in 153 per cent yield increase over control, whereas, FYM @ 5 t ha^{-1} along with combined application of lime with recommended NPK boosted the yield improvement upto 210 per cent over control i.e. farmers' practice (Table 3). Results of this study suggest that liming along with integrated nutrient management practices, if adopted properly, can lead to more than two-fold increase in groundnut productivity on acidic soils of West Tripura and other districts of the state of Tripura with similar soils. Post harvest soil analysis also showed improved status of organic C, N and P in treated plots, but available K status declined emphasising the need for close monitoring and appropriate K application in such soils (Dey and Nath, 2015).

Pod yield was significantly increased with addition of sole lime and combined application of lime and FYM along with RDF. These results are confirmatory of the previous findings of Sharma *et al.* (2006) who, based on 141 experiments in field across the Assam and Meghalaya reported 14-50 per cent increase in yield of crops in response to lime application @ 200-400 kg ha^{-1}, 22-100 per cent yield increase by recommended dose of NPK application i.e. 100 per cent NPK, and 49-390 per cent higher yield following combined use of NPK and lime compared to control i.e. farmers' practice. This might be related to the issue of balanced nutrition of crops that goes beyond the context of N, P, and K. Farmyard manure not only supplied nutrients but also improved soil conditions to produce higher yields. FYM might have stimulated the activities of micro-organisms that make the plant nutrients readily available to the crops. Use of microbial inoculants in combination with FYM favoured groundnut production.

Table 3: Effect of liming on kernel yield of groundnut under acidic soils of West Tripura

Treatments	Pod yield (kg ha^{-1})
Lime (10 % of actual LR) + RD of NPK	1130
Lime (10 % of actual LR) + RD of NPK + FYM 5 t ha^{-1}	1728
Farmer's practice	2373

iv) Adoption of integrated nutrient management (INM) practice

Integrated Nutrient Management is one of the important aspects for sustainable crop production. A field experiment was conducted during 2014-15 at the Agricultural Farm of Institute of Agriculture, Vishva Bharati, Sriniketan to study the effect of combined application of organic and inorganic fertilizers on yield, nutrient uptake, and quality content along with microbial population in Lentil-Sesame-Rice cropping system. Highest yield, nutrient uptake and quality content were obtained in integrated application of organic and inorganic fertilizer treated plots (Table 4). Micro-nutrients have a great impact on soil health in sustainable agriculture by improving growth, yield and fertility status of soils. Sulphur has a major role in case of yield, uptake and oil content of sesame (Sahu *et al.,* 2017).

Table 4: Effect of INM treatments on Sesame yield and Harvest Index

Treatments	Seed yield (kg ha^{-1})	Stover yield (kg ha^{-1})	Harvest index (%)
T1- Control(RDF::60:40:40)	485.0	1593.0	23.34
T2- $N_{40}P_{40}K_{40}$ + FYM	526.0	1660.0	24.06
T3- $N_{40}P_{40}K_{60}$+ FYM	524.0	1723.0	23.32
T4- $N_{40}P_{40}K_{80}$+ FYM	586.0	1710.0	25.52
T5- $N_{40}P_{40}K_{40}$ Mo1kg+ FYM	597.0	1720.0	25.77
T6- $N_{40}P_{40}K_{40}$ Bo0.5+ FYM	625.0	1747.0	26.35
T7- $N_{40}P_{40}K_{40}$ Zn5+ FYM	634.0	1800.0	26.05
T8- $N_{40}P_{40}K_{40}$ Zn10.5+ FYM	637.0	1880.0	25.31
T9- $N_{40}P_{40}K_{40}$ Zn21+ FYM	653.0	1950.0	25.09
T10-$N_{40}P_{40}K_{40}$+Azospirillium+ FYM	684.0	2157.0	24.08
T11- $N_{40}P_{40}K_{40}$ S_{15} + FYM	693.0	2237.0	26.17
T12-$N_{40}P_{40}K_{40}S_{30}$+FYM (Residual effect of Zn, Mo)	717.0	2283.0	26.24
T13-$N_{40}P_{40}K_{40}$ S_{45} + FYM (Residual effect of Zn, Mo)	792.0	2450.0	28.82
T14-$N_{40}P_{40}K_{40}S_{30}$+ FYM (Residual effect of Zn, Mo and B)	755.0	2307.0	26.27
T15-$N_{40}P_{40}K_{40}S_{30}$+Azospirillium+FYM (Residual effect of Zn, Mo and B)	777.0	2327.0	26.31
SEm (±)	0.064	0.227	0.271
CD (P=0.05)	0.199	0.702	0.836

v) Green manuring

Green manuring play important role in supplying nutrients for crops, improving ecological environments of agricultural fields, reducing soil erosion and pollution, restraining global warming potentials, and contributes to higher crop yields. Green manures can be fitted into rice farming systems in either the pre-rice or post-rice phase. Planting of winter green manures after the late rice has been proved to be an effective rotation pattern in improving soil conditions, soil fertility and rice yields. Many studies reported that the application of green manures changed the chemical and biological characteristics of soils. Milk-vetch (*Astragalus sinicus* L.), one kind of winter-growing legumes, has been widely used as green manure in rice fields to fertilize the soils in Japan and China (Samarajeewa *et al.*, 2005), and as an alternative N source for chemical fertilizer in double-rice cropping systems. Rape and ryegrass are also commonly used as green manures, and have the potential to improve sustainable production of double cropping rice.

vi) Pollination techniques

Planned crop pollination is considered to be the most important input as all other inputs will be just useless, if pollination does not occur. Because of yield optimizing benefits, bee pollination can play an important role in maintaining a sustainable and profitable agriculture with minimum environmental disruptions (Singh and Choudhary, 2005). According to Nagpal *et al.*, 2017, the effect of different modes of pollination on yield parameters of Indian mustard showed that the maximum number of pods plant^{-1}, pod length, pod setting (%), number of seeds pod^{-1}, thousand seed weight, seed yield plot^{-1}, seed germination (%), seed vigour and oil content (508.72 pods plant^{-1}, 5.69 cm, 86.32%, 15.66 seeds pod^{-1}, 6.87 g, 1763.0 kg ha^{-1}, 89.20%, 628.12 and 39.42%, respectively) were in open pollination followed by that in bee pollination (404.56 pods plant^{-1}, 4.92 cm, 78.33%, 14.26 seeds pod^{-1}, 6.39 g, 1557.0 kg ha^{-1}, 85.20%, 542.54 and 38.36%, respectively) and pollinators' exclusion (287.56 pods plant^{-1}, 3.89 cm, 65.87%, 12.24 seeds pod^{-1}, 5.30 g, 1301.0 kg ha^{-1}, 78.40%, 385.54 and 37.04%, respectively). Seed yield increased by 35.50 and 19.66 per cent in open pollinated and *Apis mellifera* pollinated plots, respectively as compared to pollinators' exclusion (Table 5).

Table 5: Influence of mode of pollination in *Brassica juncea* flowers on yield and oil content

Treatments	Seed yield	Oil content (%)
Bee Pollination	15.57±0.22	38.36
Open Pollination	17.63±0.28	39.42
Pollinator's exclusion	13.01±0.35	37.04
SEm (±)	0.31	0.05
CD (p=0.05)	1.04	0.18

Conclusion

Oilseed cultivation in the NEH region faces several constrains, such as water scarcity during post-monsoon season, lack of irrigation facilities, short time lag after rice harvest for seed sowing and lack of soil test based fertilizer recommendations. As a result, only mono-cropping of rice is practised and the farmers leave their land fallow. Considering oilseed Brassica as important oilseed crops for NEH region due to its lower water requirement and tolerance nature to acidity, rapeseed (*Brassica campestris* var. *toria*) after harvest of rice was introduced under zero tillage conditions. The production of oilseed under acidic soils of North-East India can be augmented by certain measures like adoption of zero tillage practice, managing phosphorus availability in acid soils quantifying right P doses, application of lime, adoption of integrated nutrient management practice, green manuring and planned crop pollination, etc.

Integrated nutrient management is one of the important issues for sustainable oil seed production. Combined use of inorganic and organic sources of nutrients is essential for maintaining soil health and optimizing crop productivity alongwith adoption of suitable cereal-oilseed cropping systems. The mulching practices, green manuring and liming are more suitable for improving oilseed production under acidic soils of North-East India. Thus, all the above techniques are recommended for the farmers of North-East India to augment oilseed production and livelihood security.

References

Choudhary, V.K., Kumar, P.S., and Bhagawati, R. 2013. Response of tillage and *in situ* moisture conservation on alteration of soil and morphophysiological differences in maize under Eastern Himalayan Region of India. *Soil Tillage Res.*, 134: 41-48.

Dey, D. and Nath, D. 2015. Assessment of effect of liming and integrated nutrient management on groundnut under acidic soil condition of West Tripura. *Asian J. Soil Sci.*, 10(1): 149-153.

Kumar, V. and Ladha, J.K. 2011. Direct seeding of rice: Recent developments and future needs. *Adv. Agron.*, 111: 297-413.

Maurya, A. and Sanjay-Swami 2017. Critical limits of available phosphorus for rapeseed (*Brassica campestris* var. toria) growing acidic soils of Meghalaya. Paper presented in the 26th National Conference of Soil Conservation Society of India at CPGS, CAU, Umiam during 11-13 September, 2017.

Mir, M.R., Mobin, M., Khan, N.A., Bhat, M.A., Lone, N.A., Bhat, K.A., Razvi, S.M., Wani, S.A., Wani, N., Akhter, S., Rashid, S., Masoodi, N.H. and Payne, W.A. 2010. Effect of fertilizers on yield characteristics of mustard (*Brassica juncea* L. Czern and Coss). *J. Phytol.*, 2(10): 20-24.

Mutetwa, M. and Mtaita, T. 2014. Effects of mulching and fertilizer sources on growth and yield of oilseeds. *J. Global Inno. Agri. Social Sci.*, 2: 102-106.

Nagpal, K., Yadav, S., Kumar, Y. and Singh, R. 2017. Effect of pollination modes on yield components in Indian mustard (*Brassica juncea* L.). *J. Oilseed Brassica*, 8(2): 187-194.

Rao, A.N., Jhonson, D.E., Sivaprasad, B. Ladha, J.K. and Mortimer, A.M. 2007. Weed management in direct seeded rice. *Adv. Agron.*,93: 153-255.

Sahu, G., Chatterjee, N., Bag, A.G. and Mukherjee, A.K. 2017. Effect of integrated nutrient management on lentil-sesame-rice cropping system in red and lateritic soils of West Bengal. *Int. J. Chem. Studies*, 5(5): 2020-2029.

Samarajeewa, K.B.D.P., Takatsugu, H. and Shinya, O. 2005. Effect of Chinese milk-vetch (*Astragalus sinicus* L.) as a cover crop on weed control, growth and yield of wheat under different tillage systems. *Plant Pro. Sci.*, 8: 79–85.

Sharma, P.D., Baruah, T.C., Maji, A.K. and Patiram, 2006. Management of acid soils in NEH Region, Natural Resource Management Division (ICAR), Krishi Anusandhan Bhawan-II, Pusa Campus, New Delhi. *Technical Bulletin*, pp. 14.

Singh, R. and Choudhary, D.K. 2005. Effect of honey bee pollination on yield and quality of mustard, *Brassica juncea* L. Paper presented in International Beekeeping Congress held on November 13-18, 2005 at Bangalore, India.

Zhao, H., Xiong, Y.C., Li, F.M., Wang, R.Y. and Qiang, S.C. 2012. Plastic film mulch for half growing season maximized WUE and yield of potato via moisture-temperature Improvement in a Semi-arid Agro ecosystem. *Agri. Water Management*, 104: 68-78.

❑❑❑

Chapter 21

Impact of Front Line Demonstrations on Blackgram (*Vigna mungo* L.) in West Garo Hills of Meghalaya

M.M. Islam*, L.K. Nath and T. Samajdar**

ICAR Research Complex for NEH Region, Umiam-793103, Meghalaya, India, * Corresponding author: E-mail: mislam01d@yahoo.co.in

** SMS (Agronomy), KVK Lakhimpur, AAU (Assam)

ABSTRACT

Two hundred forty three Front Line Demonstrations on blackgram were conducted at six different blocks of West Garo Hills district of Meghalaya on different aspect during 2010-11 to 2014-15. The yield gap between the improved technology and farmers practice ranging from 50 to 391 kg/ha. Among the blocks, the yield gap was higher in Selsella block followed by Zikzak ranging from 171 to 253 kg/ha. There is urgent need to make stronger extension services for educating the farmers in the implementation of improved technology. The technology gap of 65 to 237 kg/ha was recorded with the technology index of 6.5 to 23.7 per cent, which indicated the need to educate cultivators and lower the technology index showed the feasibility of demonstrated improved technology at location specific farmers' field. The net return (Rs 26,380/ha), additional return (Rs. 9940/ha), benefit cost ratio (3.88) and incremental benefit cost ratio (14.20) were found higher under improved technology demonstrated at Selsella block as compared to farmers practice.

Keywords: Extension gap; technology gap; technology index; yield gap.

The global pulses production was 61.50 million tonnes from an area of 70.60 million hectares with an average yield of 871 kg/ha (IIPR 2011). However, India is the largest producer of pulses in the world with 25% share in the global production. Blackgram is one of the important pulse crops contributing 10% to the total production of pulses in the country. The production and productivity of pulses in the country was 14.56 million tons and 625 kg/ha, respectively from an area of 23.63 million hectares during 2007-08. But, the annual demand of pulses is pegged at 25 million tons. The productivity of pulses in North-Eastern Region is around 800 kg/ha. Pulses are grown on marginal land under rainfed condition with low or no inputs. The area under kharif and rabi pulses in Meghalaya was 1.3 and 2.7 thousand hectares with production of 0.9 and 2.4 thousand tons and productivity of 692 and 889 kg/ha, respectively. It is mainly consumed as dal, whole or splitted, husked or unhusked. Black gram is also an integral ingredient of famous South Indian preparations like Dosa, Idli, Vada, etc. It is highly nutritive and contains high proportion of digestible protein with many essential amino acids, minerals and vitamins. Promoting pulses in fallow lands, better crop spacings, improved pulse varieties and improved market support can encourage farmers to grow more pulse. Keeping in view the low productivity of pulses, the present study was undertaken to study the impact of frontline demonstration on blackgram in five different blocks of West Garo Hills district of Meghalaya.

Materials and Methods

Two hundred forty three Front Line Demonstrations (FLDs) were conducted during 2010-11 to 2014-15 at six different blocks vi., Dadengre, Gambegre, Selsella, Betasing and Zikzak of West Garo Hills district of Meghalaya (Table 1). The objective was to transfer the improved technology to increase the productivity of blackgram through various types of demonstrations viz., varietal, improved integrated package of practices, nutrient management, soil amendment, weed management, Rhizobium inoculation and foliar application of 2% urea. All the demonstration was conducted on sandy loam soils at 0.40 to 0.50 ha area. The recommended varieties of T9 and Kalindi was demonstrated both, in broadcast and line sowing at 30x10-15cm spacing by using the seed rate of 15-20 kg/ha. The farmers of the district had apathy to use chemical fertilisers and other inputs in Blackgram cultivation. However, the demonstration of application of

DAP@20 kg /ha, soil amendment with lime @ 500 kg/ha in furrows, *Rhizobium* inoculation @20 g /kg seed and foliar spray of 2% urea were conducted in different clusters on farmers' field. The yield data for demonstrated improved technology (IT) and farmers' practice (FP) were collected from the equal areas. The demonstration was conducted to study the gaps between the potential yield and demonstration yield, extension gap and technology index. In the present evaluation study, the data on output of rapeseed cultivation were collected from demonstrated plots, besides the data on local practices commonly adopted by the farmers of the district were also collected. To estimate the technology gap, extension gap and technology index, the formulae (*Samui et al.*, 2000) has been depicted as:

$$\text{Technology gap} = \text{Potential yield} - \text{Demonstration yield}$$

$$\text{Extension gap} = \text{Demonstration yield} - \text{Farmers yield}$$

$$\text{Technology index} = \frac{\text{Potential yield} - \text{Demonstration yield}}{\text{Potential yield}} \times 100$$

$$\text{Additional return} = \text{Demonstration return} - \text{Farmers practice return}$$

$$\text{Incremental B:C ratio} = \frac{\text{Additional return}}{\text{Additional cost}}$$

To estimate the impact in terms of monetary benefits, the benefit cost ratio (BCR) analysis of improved technology (IT) over the real farmers' practices (FP) was worked out with the prevailing market value of the crop.

Results and Discussion

The data presented in Table 2 indicated that the average seed yield of blackgram varieties under varietal demonstration was 930 kg/ha as compared to 560 kg/ha in farmers practice. This indicated that the use of improved varieties contributed 66.1 per cent higher production than farmers' traditional varieties. Under improved integrated package of practices, the average seed yield of blackgram was 1050 kg/ha in improved technology, whereas that of farmers practices was 659 kg/ha with the increase of 59.33 per cent productivity over farmers practice. About 15 demonstrations were conducted on nutrient management with DAP@20kg/ha on blackgram. The average seed yield of 856 kg/ha in improved technology as compared to farmers practice (575 kg/ha). The per cent increase in yield was 48.9 over farmers practice. Fifty seven demonstration on soil amendment was conducted with lime @ 500 kg/ha in furrow application. The average seed yield of 836 kg/ha was recorded over 530 kg/ha in farmers practice with the increase in 52.0 per cent yield. Weed management demonstration was conducted only in 10 places, where a meagre per cent increase in yield was recorded. However, twenty two demonstration on *Rhizobium* inoculation with 20g/kg seed was conducted with average seed yield of 836 kg/ha over without Rhizobium inoculation (650 kg/ha), where about 28.6 per cent yield was increased over farmers practice. Dudhade *et al.*(2009) reported 9.9 per cent increased yield due to application of *Rhizobium* to chickpea seed. On the other hand, the foliar application of urea @ 2% was demonstrated at 12 different clusters, where 32.7 per cent yield of blackgram was increased as compared to non-application of foliar spray (Table 2). This indicated that the transfer of improved technology on the aspects was beneficial for increasing the farmers' income. The increasing trend due to this technology was also reported by Desai *et al.*(1999) and Poonia and Pithia, (2011). Similarly, the seed yield of blackgram was recorded highest at block Selsella (935 kg/ha) followed by Zikzak (898 kg/ha) with the increase of 42.75 and 39.22 per cent, respectively in improved technology over the traditional farmers practice (Table 3).

Yield gaps

Yield of various types of demonstration on improved technology (IT) and farmers practice (FP) of blackgram was compared to estimate the yield gaps during different years (2010-11 to 2014-15). The yield gaps presently ranging from 50 to 391 kg/ha. Under improved integrated package of practice and varietal aspect (availability of improved seeds), the yield gap was much higher (391 and 370 kg/ha, respectively), followed by soil amendment and nutrient management. This is might be due to social milieu, lack of research on pulses and effective transfer of technology. This is in conformity to that reported by Poonia and Pithia, (2011). The technology gap which corroborates to the gap in demonstration yield over potential yield ranged from 65 kg/ha at Selsella block to 237 kg/ha at Gambegre block. The technology gap observed may be attributed to dissimilarity in soil fertility status, topgraphy and weather conditions at different blocks in the district. Hence, location specific recommendations appear to be necessary to bridge the gap between the yields of blackgram. The highest extension gap was found at Selsella block followed by Zikzak block which emphasized the need to educate the farmers' to adopt improved agro-technologies to reverse the trend of wide extension gap. More and more use of HYVs by the farmers may subsequently change this alarming trend of galloping extension gaps (*Sarma et al.,* 2014). The new technologies will eventually lead to the farmers to disenchantment discontinuance of old varieties with new technology. The technology index showed the feasibility of evolved technologies at farmer's field. The lowest technology index of 6.5 per cent at Selsella block, and highest technology index of 23.7 per cent at Gambegre block, which indicated that lower the value of technology index, more is the feasibility of the technology demonstrated (Table 3).

Economic Analysis

The net return (Rs 26,380/ha), additional return (Rs. 9,940/ha), benefit cost ration (3.88) and incremental benefit cost ration (14.20) were found higher under improved technology demonstrated at Selsella block, followed by Zikzak block as compared to farmers practice (Table 4). Hence, higher benefit cost ratios proved economic viability of the improved technology (IT) demonstrated on different aspects at different blocks in West Garo Hills district of Meghalaya. This finding is in conformity with the finding of *Balai et al.,* 2012 in rapeseed and mustard and *Sharma,* 2003 in moth bean.

Conclusion

The technologies demonstrated under Front Line Demonstrations had been exploited to obtain the maximum yield, net profit and additional income of blackgram cultivation which lead to economic viability of the farming in the district. The yield and profit gaps between technology demonstrated and farmers practice might be due to biophysical, socio-economic, management, institutional, and policy factors. The highest extension gap reflects the need to educate the farmers through various means for adoption of improved technologies to reverse the trend of wide extension gap.

Table 1 : Number of Frontline Demonstrations on blackgram conducted in blocks of West

Blocks	Demonstration conducted					Total
	2010-11	2011-12	2012-13	2013-14	2014-15	
Dadengre	3	3	7	2	8	26
Gambegre	2	3	8	2	9	24
Selsella	3	4	6	3	13	29
Betasing	5	6	25	6	20	62
Zikzak	7	8	49	7	34	105
Total	**20**	**24**	**95**	**20**	**84**	**243**

Table 2 : Mean results of Frontline Demonstrations on different aspects in West Garo Hills (2010-11 to 2014-15).

Improved aspects/Varieties	No. of demonstration	Seed yield (kg/ha)			Extension Gap
		IT	FP	% increase	
Varietal T9, Kalindi	110	930	560	66.07	370
Improved integrated package of practices	17	1050	659	59.33	391
Nutrient management (DAP@ 20 kg/ ha)	15	856	575	48.87	281
Soil amendment (Lime @500 kg/ ha)	57	836	550	52.00	286
Weed management	10	710	660	7.58	50
Rhizobium inoculation	22	836	650	28.62	186
Foliar application (2% urea spray)	12	836	630	32.70	206
Total/Average	**243**	**864.86**	**612**	**42.17**	**252.86**

Table 3: Yield gaps analysis of Frontline Demonstrations on blackgram in different blocks of West Garo Hills (5 years pooled)

Blocks	Seed yield (kg/ha)			% increase	Technology gap (kg/ha)	Extension gap(kg/ha)	Technolog y index (%)
	Potential	IT	FP				
Dadengre	1000	785	590	33.05	215	195	21.5
Gambegre	1000	763	575	32.70	237	188	23.7
Selsella	1000	935	655	42.75	65	280	6.5
Betasing	1000	806	635	26.93	194	171	19.4
Zikzak	1000	898	645	39.22	102	253	10.2
Pooled	**1000**	**837**	**620**	**35.06**	**162.6**	**217.4**	**16.26**

Table 4: Economic impact of improved technology (IT) on blackgram over farmers practice (FP) in West Garo Hills, Meghalaya (5 years pooled).

Blocks	Cost of cultivation (Rs./ha)		Gross return (Rs./ha)		Net return (Rs./ha)		BCR		Additional Return (Rs/ha)	IBCR
	IT	FP	IT	FP	IT	FP	IT	FP		
Dadengre	9500	8800	29830	22420	20330	13620	3.14	2.55	6710	9.59
Gambegre	8870	8200	28994	21850	20124	13650	3.27	2.66	6474	9.66
Selsella	9150	8450	35530	24890	26380	16440	3.88	2.95	9940	14.20
Betasing	9680	9000	30628	24130	20948	15130	3.16	2.68	5818	8.56
Zikzak	9785	9055	34124	24510	24339	15455	3.49	2.71	8884	12.17
Pooled	**9397**	**8701**	**31821**	**23560**	**22424**	**14859**	**3.39**	**2.71**	**7565**	**10.87**

References

Balai, C.M., Meena, R.P., Meena, B.L. and Bairwa, R.K. 2012. Impact of front line demonstration on rapessed and mustard yield improvement. *Indian Res. J. Extn. Edn.*, 12(2): 113-116.

Desai, B.R., Patil, R.P., Patil, P.G., Kharde, P.B. and Waman, G.K. 1999. Impact of Farmers in Increasing the productivity and Area under Vijay Gram. *J. Maharashtra Agric. Uni.*, 24(1): 104-105.

Dudhade, D., Deshmukh, G.P., Harer, P.N. and Patil, J.V. 2009. Impact of Front Line Demonstrations of Chickpea in Maharashtra. *Legume Res.*, **32(3)**: 2206-208.

Poonia, T.C. and Pithia, M.S. 2011. Impact of Front Line Demonstrations of Chickpea in Gujarat. *Legume Res.*, 34(4): 304-307.

Samui, S.K., Maitra, S., Roy, D.K., Mandal, A.K. and Saha, D. 2000. Evaluation on Front Line Demonstration on Groundnut (*Arachis hypogaea* L.). *J. Indian Soc. Coastal Agri. Res.*, 18(2): 18-183.

Sarma, H., Sarma, R., Sarmah, A.K. and Upamanya, G.K. 2014. Yield Gap Analysis of Toria (Brassica campestris) in Barpeta district of Assam. *Indian Res. J. Extn. Edn.*, 14(2): 127-129.

Sharma, O.P. 2003. Moth bean yield improvement through Front Line Demonstration. *Agril. Extn. Rev.*, 15(5): 11-13.

□□□

Chapter 22

Economic Analysis of Zero Tillage in Wheat Under Rice-Wheat Production System of Haryana

Mukesh Kumar Meena

ICAR-Indian Institute of Soil and Water Conservation, RC-Koraput-763002, Odisha, [1] E-mail: lavish1986@gmail.com

ABSTRACT

Zero Tillage (ZT) has been considered as a proven technology to increase the farm income, but its adoption in the semi-arid part of western India is still uncertain. This study examines the ZT diffusion pattern (adoption and non-adoption) and estimates the farm-level impacts (changes in productivity and input use) of the ZT technology in Haryana. The most pronounced impact of ZT is at the cost-saving compartment. However, since ZT wheat farmers could sow the crop much earlier than their conventional counterparts results in gaining 8 per cent increase in yield. Factors influence wheat yield analyzed by employing production function. Late sowing of wheat under ZT practice is unwarranted, as it leads to lower productivity. Hence, extension workers should concentrate on this aspect in their training and demonstration programmes. Technology intervention needs to be complemented with policy reforms (subsidy on inputs and infrastructure) to create an enabling environment for sustainable adoption of this conservation practice. Determinants of adoption were attempted employing a multinomial logit model. Ownership of ZT drill, cropping pattern and household factors were influencing ZT adoption. Although a number of constraints for continuous adoption of ZT, including uncongenial field situation (excess moisture, undulated plots, small landholding, etc.) and difficulty to manage previous/*kharif* crop (pearl millet, cotton) residues.

Keywords: Zero tillage, Conservation, Adoption, Haryana, multinomial logit, Constraints.

Introduction

The green revolution has accelerated the production during 1965-1985. But subsequently, the increase in production did not keep pace with the population growth. The rice-wheat system is primarily irrigated, and 85 per cent of this system is concentrated in the Indo-Gangetic plain (IGP), including northern India, Pakistan, Nepal and Bangladesh (Timsina and Connor, 2001). The irrigated rice-wheat systems consumes large amount of water with the increasing competition for water from industrial, domestic and environmental sectors. Increasing water scarcity is also seen as a major contributor to stagnating productivity in the rice-wheat cropping systems in the IGP (Erenstein *et al.*, 2007). The stagnation of productivity growth in south Asia's rice-wheat system has led to increased call for conservation agriculture to meet the challenge of productivity enhancement, ensure environmental safety and conserve natural resources (Ladha *et al.*, 2003). A Rice- wheat system covers about 34 per cent of the total rice area and 42 per cent of the total wheat area in the regions which produces one-quarter of total rice and one-third of total wheat (Farooq *et al.*, 2007).

The wheat production sector of South Asia has registered only a sluggish productivity growth during the last two decades (FAOSTAT, 2015). The Indo Gangetic Plain (IGP) of India is the belt of green revolution where Rice-Wheat cropping system practised providing income and food to people. Despite the documented positive agronomic, economic and environmental impacts, conservation tillage under wheat has not yet become widely popular in many parts of the IGP. South Asia's share of the global no-tillage area is currently only marginal (Kassam *et al.*, 2014). In India, rapid and widespread adoption of Zero Tillage (ZT) has been started in Haryana state (Laxmi *et al.*, 2007) from 2001. Haryana is an important wheat-growing state in the country and produces 10.5 Mt of wheat with yield level of 4.21 tonnes per hectare (Anonymous, 2011). Farm-level impacts is one of the first steps to address the constraints in the diffusion of a production technology Hence, the present study was undertaken with the objectives of examines the ZT impact (particularly on diffusion, adoption) and estimates the farm-level impacts of the ZT technology.

Material and Methods

Zero tillage technology

Zero tillage (ZT) implies planting crops in previously unprepared field by opening narrow slots and trenches of the smallest width and depth. It is also known as zero till, no till or direct planting. ZT continues to be followed in developing countries. In this practice at least 32 per cent of the soil surface is covered by crop residues. ZT is mixture of farm management systems that involves many agricultural practices including planting, residues management, weed and pest control, harvesting and rotation (Laxmi *et al.*, 2007).

Study area and sampling

The study was carried out in three districts of Haryana *viz.*, Ambala, Kaithal and Kurukshetra. From each selected district, five villages were selected, where ZT has been practiced. Data from full adopter, partial adopters and non-adopters collected by interview method. When categories are unordered, Multinomial Logistic regression is an often-used strategy. Multinomial logit models are straight forward extension of logistic models. Suppose a decision variable (DV) has M categories. One value (typically the first, the last, or the value with the highest frequency) of the DV is designated as the reference category. The probability of membership in other categories is compared to the probability of membership in the reference category. For a DV with M categories, this requires the calculation of M-1 equations, one for each category relative to the reference category, to describe the relationship between the DV and the Independent variables (IV).

Hence, if the first category is the reference, then, form = 2… T, there will be T-1 predicted logodds, one for each category relative to the reference category. (Note that when m = 1yougetln (1) = 0 = Z1, and exp (0) = 1). When there are more than 2 groups, computing probabilities is a little more complicated than it was in logistic regression.

For the reference category, the multinomial logistic function can be expressed as below:

$$P(y_i = 1) = \frac{1}{1 + \sum_{m=2}^{T} \exp(Z_{mi})} \tag{1}$$

Then for them = 2… T, categories the equation is expressed as in equation,

$$P(y_i = m) = \frac{\exp(Z_{mi})}{1 + \sum_{m=2}^{T} \exp(Z_{mi})} \tag{2}$$

Then the multinomial logistic regression function for different categories with reference to the reference category can be expressed as per equation,

$$\ln \frac{P(y_i = m)}{P(y_i = 1)} = \alpha_m + \sum_{k=1}^{n} \beta_{mk} X_{ik} = Z_{mi} \tag{3}$$

Equation 3 provides the logits of one category with *vis-a-vis* to the reference category. It also shows the log odds of one category with the reference category is a linear function.

Adoption categories

Since the multinomial logit function takes care of categorical dependent variables of more than two viz. full adoption, partial adoption and non-adopters. The dependent variable is the ZT adoption by the households and independent variables are different socio-economic variables in the households. The determinants for full adoption and partial adoption were separately determined since the factors influence adoption of ZT varies for these two groups. The reference category of the dependent variables was non-adopters of ZT. The other categories were full adoption and partial adoption.

Result and Discussion

Socio-Economic Profile of ZT and CT Adopters

In the study area, household facilities viz. piped water, latrine, gas connection and concrete floor was more available with ZT households (Krishna *et al.*, 2012a). Average size of cultivated land higher in ZT adopter (10.19 acre) compared to non-adopter (7.66 acre)

(Kumar *et al.*, 2005) (Table 1). Level of education, large land size and household facilities exert favourable influence on ZT adoption (higher risk taking capacity/awareness about new technology). About 66 per cent ZT farmers owned tractor and 40 per cent had zero tillage drill machine (Table 3). Age has a bearing on the farmers' risk taking attitude and innovativeness in adopting new technologies. Among respondents the mean age for full adopters, partial adopters and non-adopters was 38, 42 and 45 years, respectively. The reason behind adopting new technology was percentage of young respondents, which was more in case of full adopters (Laxmi *et al.*, 2007, Kumar *et al.*, 2010).

In the study area crop production was the main source of income, growing rice during the *kharif* (June to October) season and wheat grown in the *rabi* season (November to March).The vegetables and other oilseed crops also grown in some parts. Livestock rearing and off farm income also the other activities to supplement the family income observed in study area (Table 2).

Impact of Independent Variables on ZT Adoption

The independent variable included in the adoption model covered a wide range of relatively fixed and exogenous characteristics of farm households that are expected to be associated with the ZT adoption decision. The analysis shows that variables like information provided by Govt. agency on ZT (Krishna *et al.*, 2012), ZT drill ownership (Tripathy *et al.*, 2013) and size of farm were found positively to determine the decision of farmers to choose ZT adoption. It was also observed that most of the large farmers owned ZT seed drill which had impact on ZT adoption while others have to wait to hire seed drill (Tripathy *et al.*, 2013) (Table 4). Other parameters like irrigated area by canal, share of *kharif* crop area under basmati rice, cattle population, ZT farmers in village were also having positive impact on ZT adoption in selected districts.

In the selected districts zero tillage adopters by avoiding land preparation could sow next crop in time with the use of seed drill. In study area, date of sowing of wheat crop was different for adopters and non-adopters, in case of adopters it was 16.26 days (days from Oct. 22) as compared to 20.57 days in non-adopters.Majority of ZT farmers able to sow wheat crop in first (59.38 per cent) and second week (35.94 per cent) of November and for CT farmers, it was second

(46.69 per cent) and third (17.05 per cent) week of November (Fig. 1). As research study says that delay of every successive day in planting beyond November third week decreases the grain yield (Ali *et al.*, 2010; Irfaq *et al.*, 2005, Sharma, 1992).

Economics of ZT adoption in comparison with CT adoption

The major farm inputs used for the production of wheat crop among ZT and CT adopters are mentioned in Table 5. It revealed that ZT farmers save 26.99 per cent hired labour as compared to CT. Expenditure on human labour was relatively higher in conventional method due to use of more hired labours during the plot preparation (Erenstein *et al.*, 2007). Several studies indicate that sowing of wheat by adopting ZT technology provides benefits like savings in input cost (Malik *et al.*, 2003), timely sowing, Increase in yield and higher income. In the study area, most of the farmers adopt combine harvester machine for harvesting and threshing and few adopt manual practices. Because of that there is no difference in harvesting cost (Table 6).

There was a significant difference observed in average wheat yield by adopting ZT or CT practices. Wheat yield obtained under different practices were: Full adopters (4.46 t/ha), Partial adopters (4.29 t/ha ZT, 4.29 t/ha CT), non-adopters (4.19 t/ha.) (Fig. 2). Since ZT wheat farmers could sow the crop much earlier than their conventional counterpart's results in gaining increase in yield, which is also experimental by some studies (Laxmi *et al.*, 2007, Erenstein and Laxmi, 2008, Nasrullah *et al.*, 2010).

Constraints faced by the Zero Tillage adopters

The opinion of farmers from ZT rice-wheat production system was collected regarding the major constraints faced by them in cultivation of ZT wheat. The constraints were pooled into some groups based on field/plot characteristics, cropping pattern, undesirable impacts of technology, scarcity of the ZT drill, etc.

Field/plot characteristics include excess moisture in field, soil structure/ undulation of land unsuitable for ZT, lack of moisture during sowing and small land holding. Out of these field constraints excess moisture in field was major reason for 14 per cent discontinuous adoption. In case of non-adopters, the major reason was soil structure/ undulation of land (14 %) or small land holding (11%). Only 8 per cent adopters faced the field/plot problem during the adoption.

Undesirable impacts of technology include high aphid infestation, negative impacts on soil quality and structure, high fertilizer requirement, high weed infestation, high irrigation requirement, low yield and low germination percentage. Out of these undesirable impacts, the major reason was high aphid infestation (19%) and low yield (22%) for discontinuous adoption.

Cropping pattern constraints include difficulty to manage kharif crop (pearl millet, cotton, rice) and residues and early harvesting of kharif crops. Majority (54%) of non-adopters did not adopt ZT due to these constraints. Similarly, this problem also faced by adopters (9%). Scarcity of ZT drill was the cause of non-adopters (14%).

Conclusion

This study examines the ZT diffusion pattern (adoption and non-adoption) and estimates the farm-level impacts (changes in productivity and input use) of the ZT technology in Haryana. The most pronounced impact of ZT is at the cost-saving compartment because of ZT wheat farmers could sow the crop much earlier than their conventional counterpart and early sowing is associated with higher yield, a significant and positive yield impact (Increased by 8 per cent) observed in the study area. By avoiding land preparation, cost of cultivation could be reduced by the ZT farmers. Despite this, there exists a number of constraints for continuous adoption of ZT, including uncongenial field situation (excess moisture, undulated plots, small landholding, etc.), and difficulty to manage previous/kharif crop (pearl millet, cotton, rice) residues. Since this technology is also effective in reducing the farmer dependency on external inputs and ensures sustainable production of wheat, acreage under ZT should be further expanded, possibly through innovations that facilitate ZT operation in the adverse plot scenarios, especially in the cotton-wheat and pearl millet-wheat cropping systems. The recent ZT technological advancements that overcome many of the afore-mentioned constraints should be popularized among ZT adopters.

Acknoledgements

The first author acknowledges the financial support provided by the CIMMYT, New Delhi in terms of contingency grants for data collection. A special thanks to Dr. Vijesh V Krishna for his valuable suggestion, creative comments and help in data collection throughout study period.

Table 1: Socio-economic profile of ZT adopters and non-adopters

Characteristics	Adopters (N= 94)	Non-adopters (N= 40)
No. of adults in the household	4.97 (1.80)	5.28 (1.81)
Cultivated land owned by the household (Acres)	10.19 (9.86)	7.66 (4.76)
Education of head of households (Schooling years)	7.13 (3.17)	6.76 (3.04)
Piped drinking water (per cent)	96	77
Gas connection (per cent)	79	70
Concrete floor for house (per cent)	64	47
Toilet (per cent)	91	77

Table 2: Source of family income in study area (per cent)

Particulars	ZT (adopters)	CT(non adopters)
Crop production	82.5	80.0
Livestock	11.0	7.00
Services	4.00	8.00
Business	2.00	4.00
Others	2.00	1.00

Table 3: ZT drill and tractor ownership by adoption category (per cent)

	ZT adopters (n=94)	Partial adopters (n=40)	Non-adopters (n=46)	Overall (n= 180)
Tractor	66.00	32.00	24.00	47.77
Zero tillage drill	40.00	25.00	0	26.66

Table 4: Factor affecting ZT adopters- results of multinomial logit model

Variable	Coefficient (standard error)			
	Full adoption over non- adoption	Exp. (b)	Partial adoption over non- adoption	Exp. (b)
Land owned by households	0.432 (0.317)	1.540	0.500 (0.3350)	1.649
ZT drill ownership	3.664 (1.280)	39.023	3.928 (1.258)	50.787
Number of rooms per HH member	0.650 (-0.746)	1.916	1.157 (0.766)	3.181
Telephone expenditure	0.003 (-0.002)	1.003	0.001 (0.002)	1.001
Total amount of credit taken by the hh	-2.26E-06 (1.49E-06)	1.000	-2.37E-07 (1.15E-06)	1.000
HOHH: education	-0.065 (0.046)	0.937	-0.0386 (0.0499)	0.962
Area irrigated by canal	0.029 (0.018)	1.029	0.0188 (0.0191)	1.019
Share of kharif crop area under Basmati rice	0.022 (0.007)	1.002	0.0019 (0.0079)	1.002
Household size	-0.368 (0.187)	0.692	-0.0091 (0.1861)	0.991
No. of cattle's	0.220 (0.163)	1.246	0.1489 (0.1582)	1.161
Govt.	0.443 (0.282)	1.558	0.2935 (0.272)	1.341
ZT adoption in the village	0.044 (0.011)	1.045	0.0020 (0.0123)	1.002
Kaithal farmers	-1.222 (0.683)	0,295	0.5092 (0.733)	1.664
Ambala farmers	0.547 (0.625)	1.728	0.9778 (0.761)	2.141
Intercept	-1.988 (1.489)		-2.219 (1.541)	
Number of observations			180	
R^2			0.30	
Chi^2			114.3	
Log likelihood			-128.205	

Table 5 : Input use structures in wheat production

Variable	Mean			Percent difference
	Unit	ZT	CT	
Seed	Kg/ha.	107	109	-1.26
Fertilizer	Kg/ha.	370	362	2.21
weedicide	g/ha	1040	890	16.85
(i) family labour (imputed)	Number	4.68	4.75	-1.44
(ii) hired labour	Number	3.75	5.14	-26.99
Land preparation	Number	0	5.08	-100.00

Table 6: Cost structure of wheat production (Rs/ha.)

Variables	Mean		Percent differences
	ZT	CT	
Seed	1932	1993	-3.06
Fertilizer	4100	3600	13.89
weedicide	1800	1750	2.86
(i) Family labour (imputed)	1921	2002	-4.05
(ii) hired labour	1739	2039	-14.71
(i) Land preparation	0	2727	-100.00
(ii) Harvesting	1847	1874	-1.44
Total paid out cost	15055	16032	-6.09

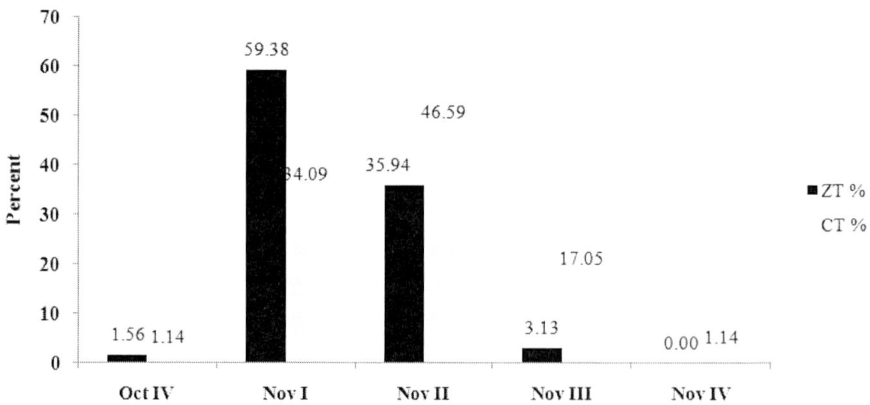

Fig. 1: Wheat sowing time of ZT adopters and non-adopters (CT)

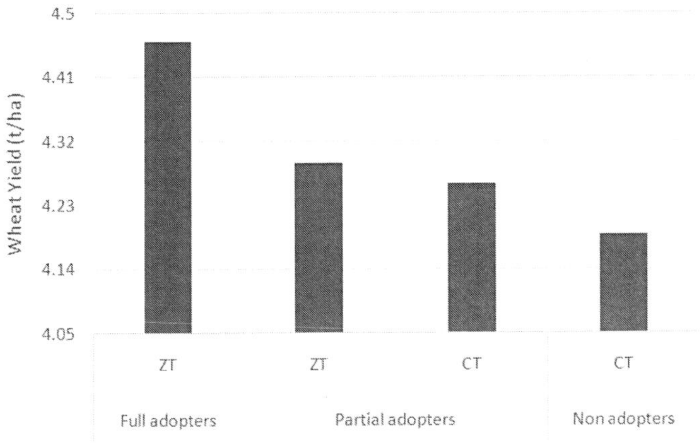

Fig. 2: Wheat yield under ZT and CT

References

Ali, M.A., Ali, M. and Sattar, M. 2010. Sowing date effect on yield of different wheat varieties. *Journal of Agriculture Research*, 48(2): 157-162.

Anonymous, 2011. Economic Survey of Haryana 2011-12, Department of Economic and Statistical Analysis, Yojana Bhawan, Panchkula, Haryana.

Erenstein, O., Malik, R.K., and Singh, S. 2007, *adoption and impacts of zero tillage in the irrigated rice-wheat system of Haryana.* (Report), CIMMYT, New Delhi.

Erenstein, O., Laxmi, V. 2008. Zero tillage impacts in India's rice-wheat systems: A Review. Soil and Tillage Research 100(1-2): 1–14. 10.1016/j.still.2008.05.001.

FAOSTAT, 2015. Food and Agriculture Organization of the United Nations, FAOSTAT database ("Crops"), Rome. http://faostat3.fao.org/home/E. Accessed 1 November, 2015.

Farooq, M., Siddique, K.H.M. 2015. Conservation agriculture: Concepts, brief history, and impacts on agricultural systems. Springer International Publishing, Heidelberg.

Irfaq, M., Muhammad, T., Amin, M. and Jabbar, A. 2005. Performance of yield and other agronomic characteristics of four wheat genotypes under natural heat stress, International Journal of Botany, 1(2): 124-127.

Kassam, A., Derpsch, R., Friedrich, T. 2014. Global achievements in soil and water conservation: The case of conservation agriculture.

International Soil and Water Conservation Research 2(1): 5–13. 10.1016/ S2095-6339(15)30009-5.

Krishna, V.V., Aravindakshan, S., Chowdhury, A., Rudra, B. 2012a. Farmer access and differential impacts of zero tillage technology in the subsistence wheat farming systems of West Bengal, India, Mexico DF.

Kumar, A., Prasad, K., Kushwaha, R.R., And Rajput, M.M. 2010, Determinants influencing the acceptance of resource conservation technology: Case of zero tillage in rice-wheat farming systems in Uttar Pradesh, Bihar and Haryana. Indian Journal of Agricultural Economics, 65(3): 448-460.

Kumar, U., Gautam, U.S., Singh, S.S., Subhash, N., Singh, K., Kumar, R., 2005. Zero tillage technology in wheat cultivation, Proceedings of the Project Workshop on accelerating the adoption of resource conservation technologies in rice-wheat systems of the Indo-gangetic plains, Hisar, India. pp. 239-244.

Ladha, J.K., Hill, J.E., Duxbury, J.M., Gupta, R.K. and Buresh, R.J. 2003, Improving the productivity and sustainability of rice-wheat system: Issue and impacts. ASA special publication no. 65.

Laxmi, V., Erenstein, O. and Gupta, R.K. 2007, Impact of zero tillage in India's Rice-wheat system, CIMMYT, Mexico.

Malik, R.K., Yadav, A., Singh, S., Sardana, P.K., Gill, G., Hobbs P.R. and Bellinder, R. 2003. Herbicide resistance management and introduction of zero tillage in wheat in India. Proceedings of Weed Science Society of America, 43-55.

Nasrullah, M.H., M.S. Cheema and M. Akhtar, 2010. Efficacy of different dry sowing methods to enhance wheat yield under Cotton-Wheat cropping system. Crop and Env. 1(1): 27-30.

Sharma, R.C. 1992. Duration of the vegetative and reproductive period in relation to yield performance of spring wheat. European Journal of Agronomy, 1: 133-137.

Timsina J. and Connor D.J. 2001. Field crop research, 69: 93-132.

Tripathi R., Raju, R. and K. Thimmappa, 2013. Impact of Zero Tillage on Economics of Wheat Production in Haryana *Agricultural Economics Research Review* Vol. 26(No.1) January-June 2013 pp. 101-108.

Vijay Laxmi and Mishra, Vinod, 2007, Factors Affecting the Adoption of Resource Conservation Technology: Case of Zero Tillage in Rice-Wheat Farming Systems. *Indian Journal of Agricultural Economics*, 62(1): 126-136.

□□□

Chapter 23

Farmers Participatory Balance Nutrient Management for Sustaining Productivity and Income: A Case Study from Semi-Arid Bundelkhand

D.R. Palsaniya

ICAR - Indian Grassland and Fodder Research Institute, Jhansi–284003 (U.P.)
Senior Scientist and Corresponding author, E-mail: drpalsaniya@gmail.com)

ABSTRACT

Participatory trials on balance nutrients were conducted during 2009-2012 in a severely micro-nutrient deficit semi-arid central Indian watershed to promote balance use of nutrients. Application of micro-nutrient increased groundnut pods per plant by 9%, seeds per pod by 6% and pod yield by 13% (1132 kg ha^{-1}) compared to control in entry point trials. Similarly, chickpea variety JG-130 grown with balance fertilizers also recorded 15 and 40% higher grain yield compared to JG-130 and local chickpea variety, respectively grown with traditional practices. Application of B, Zn and S increased mean groundnut pod yield by 16.8% over control in further upscaling trials. While investigating residual effect of balance fertilization, it was observed that grain yield of succeeding wheat, barley, chickpea and mustard increased by 8.3, 10.0, 8.8 and 9.2%, respectively. Further, the sustainability index of direct balance fertilization to groundnut was 0.8 and it was 0.76, 0.74, 0.76 and 0.83 in residual effect in succeeding wheat, barley, gram and mustard, respectively. Although nearly 88% farmers were convinced by the impact of micro-nutrients, but only 9% farmers actually used them in *kharif* 2012 and availability of micro-nutrients was emerged as the major hurdle in upscaling.

Keywords: Adoption, Balance nutrient, Micronutrient, Participatory trials, Watershed.

1. Introduction

Soil fertility crisis is being viewed as the biggest threat to agricultural sustainability (Takkar, 1996). A big challenge for agricultural scientists, farmers and planners over the coming decades will be meeting world's increasing food demand from degrading natural resource base (Palsaniya *et al.*, 2012a). The practice of intensive agriculture in terms of raising more crops per unit time and over use of high analysis fertilizers has resulted in faster depletion of secondary and micro-nutrient reserves from Indian soils (Vasuki, 2010; Palsaniya and Ahlawat, 2007). Soil surveys in semi-arid regions of India have consistently shown multi-nutrient (especially N, P, K, S, B and Zn) deficiencies due to continuous cropping and limited use of nutrient inputs (Reddy *et al.*, 2005). Analysis of 3,00,000 soil samples from across the country indicated that the average deficiency of Zn, Fe, Mn and Cu in Indian soils was 49, 12, 4 and 3%, respectively (Shukla *et al.*, 2012). Similarly, analysis of 50,000 samples showed B and Mo deficiencies in 33 and 12% of the soil samples, respectively. Further, it is estimated that approximately 68 m ha area (49% of cropped land) is suffering from S deficiency in India (The Financial Express, 2006). None of the farmers in Bundelkhand are applying micro and secondary nutrients to their field crops (Palsaniya *et al.*, 2008, 2009b). Instead of single nutrient deficiency, multi-nutrient deficiencies are emerging fast in vast areas of India and not only affecting the crop productivity but also creating malnutrition and health related problems (Vasuki, 2010; Rattan *et al.*, 2009). The deficiency of micro-nutrients is becoming a limiting factor for sustainable productivity in many soils in spite of application of recommended dose of NPK (Swarup, 1998) and therefore, as long as agriculture remains a soil-based industry, sustainable increase in farm productivity is not possible without ensuring adequate and balance supply of nutrients to crops.

Enhanced productivity due to balance fertilizer application (including micro-nutrient) has widely been reported from across the country and for wide variety of crops (Kokate *et al.*, 2012; Shukla and Behera, 2012; Palsaniya and Ahlawat, 2009; Saxena, 1995; Rego *et al.*, 2005; Vasuki, 2010; Rattan *et al.*, 2009). Saxena (1995) observed that balance fertilization is necessary for sustainability over time. He showed that wheat yields became uneconomical after 5 years, when only N fertilizer is applied. Even annual field applications of NP and NPK fertilizers were insufficient to sustain yields over the long term and when balanced fertilizers were applied yields increased and fields remained productive despite continuous cultivation. Shukla and Behera (2012) provided a critical review of response of various crops to micro-nutrient fertilizer use. They reported 160 to 513 kg ha^{-1} increase in average productivity level due to Zn application and opined that proper management of Zn alone could contribute 18.5 million tons to the total food grain production. Similarly, the average increase in productivity of pulses, oilseeds and cereals reported across the country were 34.5, 19.5 and 16.5% higher due to B application over the NPK. Recently, Kokate *et al.* (2010 and 2012) had brought out in sharp focus the advantages in increasing the productivity of various crops, such as cereal, coarse cereal, oilseeds, pulses, vegetable and food crops due to balance fertilizer application from the data generated by the Krishi Vigyan Kendras (KVKs) under on-farm conditions from various parts of India. Palsaniya and Ahlawat (2009) reported that application of 30 kg S ha^{-1} to pigeonpea increased its grain yield by 470 kg ha^{-1} compared to control. While investigating direct and residual effect of S in pigeonpea-wheat cropping system, they further revealed that S applied to pigeonpea at 60 kg ha^{-1} left behind 1.19 kg ha^{-1} more S in soil than 30 kg ha^{-1}, which resulted in higher grain yield of succeeding wheat by 580 kg ha^{-1}.

It is repeatedly being emphasized that the current adoption rate of improved agro-techniques by Indian farmers is very poor. As an alternative, Farmers Participatory Research, which was started in 1980s, emphasizes on-farm research with active participation of farmers so that these agrotechniques may readily be adopted. The watershed projects in recent times are being implemented in participatory mode as per the New Common Guidelines for Watershed Development Projects in India-2008. Participatory integrated watershed management has been a prominent approach not only for resource conservation, livelihood security and agricultural sustainability (Palsaniya *et al.*, 2012a, 2012b, 2011; Samra, 1997; Wani *et al.*, 2007) but also being used as a vehicle for transfer of improved agro-techniques to farmers (Palsaniya *et al.*, 2012b, 2010). The work done and experience gained on participatory balance fertilization and micro-nutrient use in such a model watershed project is being presented as a case study in this paper.

2. Materials and Methods

Study Site

The present study was conducted at Domagor-Pahuj watershed located in Babina block of Jhansi (Uttar Pradesh). This watershed is located between 25° 28′ 16″ to 25° 31′ North Latitude and 78° 25′ to 78° 28′ and 8″ East Longitude. The treatable area of watershed is 1373 ha. The watershed comprises of three villages, namely Domagor, Dhikauli and Nayakhera. Detailed baseline survey indicated that the main source of livelihood is agriculture, animal husbandry and farm labour. Majority of the farmers are small and marginal. The main Kharif crops are groundnut, blackgram, greengram, sesame, sorghum and maize while major Rabi crops are wheat, chickpea, mustard and lentil. The productivity of crops is low as compared to other regions of the state. Open dug wells are the main source of irrigation. The agricultural land is sloppy and has multi-directional complex slopes.

Initial Soil Fertility Status

The pre-dominant soil order is Alfisol (red soil) with some black soil patches in low lying areas. To ascertain the soil fertility status of watershed, 80 representative soil samples were drawn, processed and analyzed at International Crop Research Institute for Semi-Arid Tropics (ICRISAT), Hyderabad. The pH and electrical conductivity (EC) varied from 7.0 to 8.5 and 0.04 to 0.50 dSm^{-1}, respectively. 50% of soil samples had organic carbon less than 0.5%. Nitrogen, phosphorus and potassium were low in 56, 18 and 25% samples, respectively. Alarmingly, Zn, B and S were deficit in 58, 99 and 91% soil sapmles, respectively in the watershed area.

Methodology

Realizing the widespread deficiency of micro-nutrients in Domagor-Pahuj watershed it was decided to conduct micro-nutrient based participatory trials as entry point activity (EPA) to promote balanced fertilization. Accordingly, two groundnut varieties ICGV-91114 and ICGS-44 along with local Jhumku during *kharif*, 2009 and chickpea variety JG-130 during *rabi* 2009-10 were grown with and without micro-nutrients in on-farm trials in participatory mode. In case of groundnut, each one acre field of farmer was divided into six blocks of equal size and above three varieties were grown with and without micro-nutrients. Similarly, in case of chickpea, the farmer's field was divided into three blocks and JG-130 was grown with and without micro-nutrients and also compared with local variety grown with traditional package of practice. The number of trials in groundnut and chickpea were nine and sixty, respectively. Further, inspired by the impact of micro-nutrients in above trials, twenty nine more participatory trials were conducted with local groundnut variety during *kharif*-2011 for upscaling and to promote balanced fertilization and their residual effect was monitored in subsequent crops (wheat, barley, chickpea and mustard) grown during *rabi*-2011-12. Local varieties of above crops were raised (except barley; Narendra–2 variety) with traditional package of practices.

Groundnut was raised as per standard package of practices and applied with recommended dose of fertilizers, gypsum @ 200 kg ha^{-1}, zinc sulphate @ 25 kg ha^{-1} and Agribor (for B) @ 2.5 kg ha^{-1}. Disodium Octaborate Tetrahydrate ($Na_2B_8O_{13}.4H_2O$) is marketed in the trade name of Agribor and contains 20% Boron while Zinc Sulphate (Agriculture grade $ZnSO_4$) contains 21% Zn and 10% S. They were applied directly to soil through broadcasting during field preparation before sowing. Groundnut was sown in second week of July and last week of June in 2009 and 2011, respectively and harvested during third week of October during both the years. In case of chickpea, JG-130 was sown in second week of October, 2009. The 2009-10 crops of groundnut and chickpea were taken as rainfed while 2011-12 crops were taken as irrigated (groundnut, 2-3 irrigations; wheat, 4-5 irrigations; barley, 1-2 irrigations; chickpea, 1 irrigation and mustard, 1 irrigation). All the participatory trials were of one acre each.

To monitor the impact of balanced fertilization (micro-nutrient), uniform crop sampling was done. Five samples each of 3m x 3m size were taken as per treatment (with and without micro-nutrients and traditional practice) from each trial at harvest and biometric observations were recorded for each crop. The Sustainability Index (SI) of balanced fertilization was calculated using following formula. SI = (Y_{mean} – Standard Deviation)/ Y_{max}, where Y_{mean} is the mean yield of crop and Y_{max} is maximum yield. In order to know the adoption behaviour and hurdles in adoption of micro-nutrient use, 80 farmers from the watershed were interviewed through well structured questionnaire. They included participating and non participating farmers as well.

3. Results and Discussion

Micro-nutrient Based Entry Point Activities (EPAs)

The analysis of soil showed widespread deficiency of zinc, sulphur and boron in the watershed area. Therefore, it was decided to conduct micro-nutrient based EPAs in groundnut and chickpea in the watershed. Application of balanced fertilizers (RDF + micro-nutrients + S) significantly enhanced growth, yield attributes and yield of groundnut in participatory trials during *kharif*, 2009 (Table 1). Micro-nutrient increased groundnut pods/plant by 9%, seeds/pod by 6%, 1,000-grain weight by 4%, pod yield by 13%, haulm yield by 14%, biological yield by 13% and shelling by 2% compared to crop raised without micro-nutrient. The local variety Jhumku performed better as compared to the newly introduced varieties. Jhumku recorded 26.4 and 21.3% higher pod yield over ICGV-91114 and ICGS-44, respectively. This might be due to the fact that these new varieties started 10-15 days early flowering than Jhumku and there was a dry spell of 21 days during this period, which resulted into poor flowering and pegging in these varieties. Rego *et al.* (2007) also reported that the percentage increase in pod yield of groundnut due to the application of SBZn over the farmers practice were 33, 48, and 29 during 2002, 2003 and 2004, respectively in participatory trials.

The Chickpea variety JG-130 grown with balanced fertilizers (RDF + micro-nutrients + bio-fertilizers) under rainfed condition also recorded 15 and 40% higher grain yield, respectively compared to JG-130 and local chickpea variety grown with traditional practice without micro-nutrients (Fig. 1). This showed the potential of micro-nutrients even under rainfed condition. Further, farmers can get 22% higher grain yield simply by growing improved high yielding chickpea variety like JG-130 compared to local variety. Rego *et al.*, (2005 and 2007) also showed similar encouraging yield increase in these crops from watershed based participatory trials conducted in Andhra Pradesh, Maharashtra, Karnataka and Rajasthan.

Upscaling Micro-nutrient Use

To promote further use of micro-nutrients, twenty nine more participatory trials were conducted during *kharif*-2011 under irrigated conditions in groundnut and residual effect was monitored in subsequent crops grown during *rabi*-2011-12 (Table 2 and 3). Twenty trials were monitored for data collection. It is worthwhile to mention that water resources for the purpose were developed within watershed through conservation activity. Agribor (B), Zinc Sulphate (Zn) and Gypsum (S) were provided at 50% cost from the project and rest of the inputs were managed by farmers themselves. The average productivity of groundnut with balanced fertilization under these participatory trials was 1842 kg pods ha^{-1} with range of 1625 to 2120 kg ha^{-1}. However, mean groundnut pod yield without micro-nutrients and S was 1577 kg ha^{-1} with range of 1370 - 1895 kg ha^{-1}. It was observed that application of B, Zn and S increased the groundnut pod yield by 16.8% over control. The range of response of groundnut to balanced fertilization was 5 to 33%. The Sustainability Index (SI) of balanced fertilization to groundnut was as high as 0.8.

The beneficial effects of balanced fertilization may be attributed to the better growth and productivity of crops, which resulted in better quality, lower production costs, better profitability, improved chances for producing a good yield under adverse climatic and soil conditions, reduced incidence of plant diseases and pests, safeguards to natural resources, preservation of soil fertility for future generations by increasing water, nutrient, energy and land-use efficiency and environmental benefits through reduced leakages into the biosphere via leaching, denitrification and volatilization losses (Kokate *et al.*, 2012; Shukla and Behera, 2012; Palsaniya and Ahlawat, 2009; Rego *et al.*, 2005; Vasuki, 2010; Rattan *et al.*, 2009). The Law of Minimum clearly indicate the significance of balanced supply of all the nutrients for maximum yield and yield level cannot be further raised, if one of the essential nutrient becomes limiting factor even though others are at optimum (Tisdale *et al.*,1985). The similar enhanced productivity and profitability due to balanced fertilization was also reported on larger scale by Kokate *et al.*, (2012), Shukla and Behera, (2012) and Rego *et al.*, (2005) from across the country.

Residual Effect

The residual effect of balanced fertilization was monitored in subsequent crops grown on above plots during *rabi*-2011-12 and it was observed that application of B, Zn and S to groundnut had residual effect on succeeding wheat, barley, chickpea and mustard (Table 3). The mean yield of wheat was 3010 kg ha^{-1} when balanced nutrients were applied to the preceding groundnut crop. It was 231 kg higher compared to wheat yield (2779 kg ha^{-1}) obtained from the field where no B, Zn and S were applied to preceding groundnut. Similarly, average barley grain yield also increased by 337 kg due to residual effect of balanced nutrients applied to preceding groundnut. The mean chickpea yield under residual plots (1783 kg ha^{-1}) was 145 kg higher compared to yield obtained from control plots (1638 kg ha^{-1}). Likewise, average mustard yield under residual plots (1385 kg ha^{-1}) was 116 kg higher compared to yield obtained from control plots (1269 kg ha^{-1}). The grain yield of succeeding wheat, barley, chickpea and mustard increased by 8.3, 10, 8.8 and 9.2%, respectively, when balanced nutrients were applied to the preceding groundnut crop compared to yield obtained under control. The Sustainability Index (SI) of residual effect of balanced fertilization in wheat, barley, gram and mustard was 0.76, 0.74, 0.76 and 0.83, respectively.

Economic Gain

Besides harvesting more grain yield, farmers also benefitted economically by spending only Rs. 1750/ha for B, Zn and S over and above other nutrient inputs. The net economic gain due to direct application of B, Zn and S to chickpea and groundnut were Rs. 3024 and 7155 ha^{-1}, respectively (Fig. 2). The economic gain due to residual effect of B, Zn and S applied to previous crop was also recorded in succeeding mustard, wheat, barley and chickpea to the tune of Rs. 2565, 2772, 3303 and 4060 ha^{-1}, respectively. In a similar study, Rego *et al.,* (2005) also reported net economic gain of Rs. 8200 ha^{-1} for maize and Rs. 2900 ha^{-1} for pigeonpea in Mahbubnagar watersheds. Similarly, in Nalgonda watersheds, net economic gain for castor was Rs. 1600 ha^{-1} while that for Mung bean was Rs. 2700 ha^{-1}. In Kurnool, the net economic gain for groundnut was Rs. 6500 ha^{-1} while that for pigeonpea was 3200 ha^{-1} due to balanced fertilization. Balanced application of micro-nutrients with major nutrients to different crops showed similar response in watershed projects in other parts of the SAT in India. Participatory work at watershed projects in Vidisha, Guna and Dewas districts of Madhya Pradesh revealed 40 to 83% increase in groundnut grain yield with considerable residual effect on yield of succeeding chickpea and wheat (Rego *et al.,* 2005). Similarly, at Lalatora watershed (Vidisha, Madhya Pradesh), they further reported that combined application of B and S to soyabean increased its grain yield by 26, 18 and 53% in 2000, 2001 and 2002, respectively over control (farmer input).

The residual effect of these micro-nutrients in the following wheat crop was in the range of 5 to 18%. Further, on average (the average of all on-farm trials on soyabean-wheat cropping system), the soyabean grain yield for combined application of B +S treatment was 1770 kg ha^{-1} while it was only 1400 kg ha^{-1} in control. Grain yield of the following wheat crop was 3700 kg ha^{-1} in treated plots while control plot yielded 2700 kg ha^{-1} of wheat grain. The net profit from soyabean-wheat system was Rs. 26,450 ha^{-1} in treated plots compared to Rs. 17,760 ha^{-1} in control plot. A similar research work at Adarsha watershed (Andhra Pradesh) also revealed that application of B and S increased sorghum grain yield from 13 to 29% while increase in maize grain yield was 20 to 39% over control. ICRISAT while scaling–up the watershed interventions for enhancing agricultural productivity in Andhra Pradesh during 2003 under APRLP selected total 50 watersheds and reported an increased response in yield of various crops to application of S, Zn and B along with N and P. The response obtained by combined application of B, Zn and S at farmer level and at optimum N and P level was (51% and 76% in maize, 41% and 61% in sorghum, 47% and 78% in groundnut, 41% and 61% in Mungbean, 71% and 90% in pigeonpea and 54% and 70% in castor, respectively (Rego *et al.*, 2005, 2007). Combined application of B and S with other nutrients to maize in Bundi watershed (Rajasthan) increased grain yield by 60% over control (Rego *et al.*, 2005). In other 279 trials, conducted in Banswara, Jhalawar and Tonk districts of Rajasthan in maize, soybean, pearlmillet and groundnut crops during 2010 revealed that combined application of Zn + B + S increased yield of these crops in range of 10 to 50% as compared to the farmers practice (Girish Chander *et al.*, 2011).

Spread of Micro-nutrient Use and Hurdles in Adoption

Investigations were made on spread of micro-nutrients and hurdles in adoption through structured Questionnaire Proforma and its outcome is summarized in Table 4. It was learnt that despite concerted efforts, spread and adoption of micro-nutrients (balanced fertilization) was not satisfactory. Although nearly 88% farmers were convinced by the impact of micro-nutrients, but only 9% farmers actually used them in *kharif*-2012. Nearly 76% farmers wanted to use micro-nutrients but only 28 tried to buy them. Availability and awareness found to be the major hurdles in spread of micro-nutrient use as 68% farmers mentioned that these are not easily available and

10% cited lack of knowledge regarding micro-nutrients. The per cent of farmers not convinced by the quality, cost and effectiveness of micro-nutrients was 8, 5 and 10%, respectively. Therefore, there exist a greater scope for micro-nutrient manufacturing and marketing companies to make available quality micro-nutrient fertilizers at affordable cost as 76% of the farmers surveyed wanted to use them. More farmers may be persuaded to adopt micro-nutrients through spreading awareness regarding effects of micro-nutrients on crop productivity.

4. Conclusions

A widespread deficiency of S, Zn and B was observed and balanced application improved the yield of various crops in range of 13-17% in participatory entry point trials. Further, the residual effect of balanced fertilization increased the grain yield of succeeding crops by 8-10%. Nearly 88% farmers appreciated the impact of balanced fertilization, but only 9% farmers actually used them in succeeding season and availability of micro-nutrients was emerged as the major hurdle in upscaling. Therefore, researchers, extension personnel, policy-makers, fertilizer industry and dealers, all have to contribute to make micro-nutrients available to farmers at affordable prices.

Acknowledgement

The author is thankful to the farmers of watershed area for their active co-operation and participation during the study and to the Ministry of Agriculture, Government of India for financial support for the watershed project.

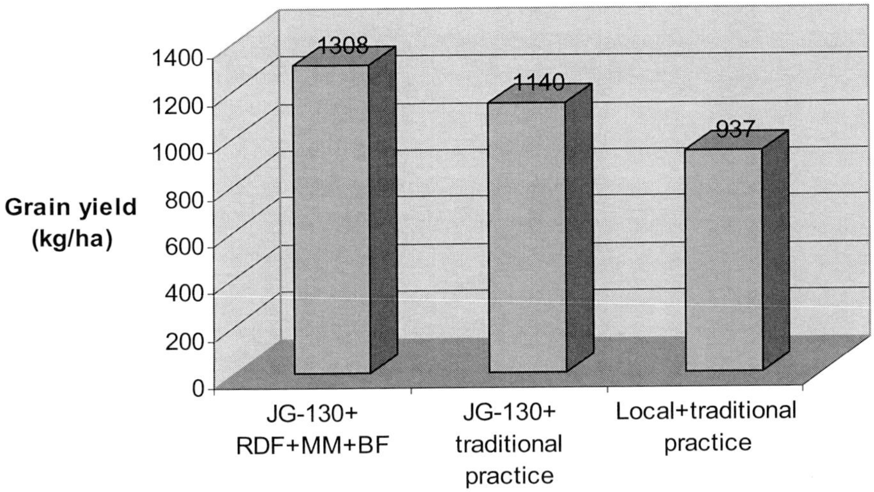

Fig. 1: Performance of chickpea (*rabi*, 2009-10)

Where RDF = recommended dose of fertilizer, MM = micro-nutrient mixture @ 12.5 kg/ha (containing 6% zinc, 3% Fe, 1.5% Mn and 0.5% Ca) and BF = Bio-fertilizer - Rhizobium and PSB

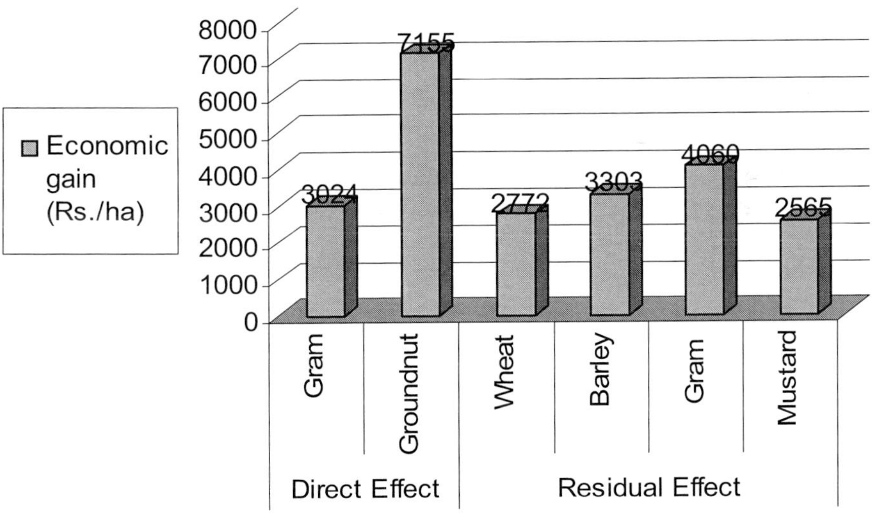

Fig. 2: Net economic gain due to direct and residual effect of B, Zn and S application

Table 1: Groundnut varieties with micro-nutrients in on farm trials

Treatment	Pods $Plant^{-1}$	Seeds Pod^{-1}	1000-Seed weight (g)	Pod yield (kg ha^{-1})	Haulm yield (kg ha^{-1})	Biological yield (kg ha^{-1})	Harvest index (%)	Shelling %
Variety								
ICGV-91114	10.2	1.81	330.5	969	2130	3099	31.3	67.2
ICGS-44	11.0	1.94	335.8	1010	2313	3323	30.4	68.3
Jhumku	12.4	2.10	345.2	1225	2793	4018	30.5	65.8
SEm	0.4	0.05	3.3	68.4	154	224	0.4	0.45
CD at 5 %	1.2	0.14	9.7	202	452	658	1.0	1.3
Micro-nutrient								
With*	11.7	2.01	343.2	1132	2566	3698	30.6	67.7
Without	10.7	1.89	331.2	1005	2257	3262	30.8	66.5
SEm	0.3	0.03	4.1	40.8	99	142	0.2	0.4
CD at 5 %	0.9	0.10	10.2	119	289	414	NS	1.1

*Gypsum @ 200 kg ha^{-1} + Zinc Sulphate @ 25(kg ha^{-1} + Agribor @ 2.5 kg ha^{-1}

Table 2: Groundnut yield with micro-nutrients in participatory trials

Name of farmer	Pod yield (kg ha^{-1})		% yield increase
	With micro-nutrients (B, Zn) & S	Without micro-nutrients & S	
Shri Ram Dayal	1850	1450	27.6
Shri Atma Ram	1980	1660	19.3
Shri Arun Kumar	1790	1525	17.4
Shri Kallu Sahu	1900	1630	16.6
Shri Bahadur Singh	2020	1515	33.3
Shri Daya Ram	1660	1370	21.2
Shri Sharda Saran	1710	1370	24.8
Shri Ram Pal Singh	1800	1456	23.6
Shri Arvind Singh	1625	1420	14.4
Shri Puran Rai	1810	1600	13.1
Shri Jindel Singh	1720	1560	10.3
Srimati Kasturi Devi	1925	1690	13.9
Shri Vijay Singh	1930	1710	12.9
Shri Om Prakash	1695	1450	16.9
Shri Jagat Ram	1870	1690	10.7
Shri Tula Ram	1820	1695	7.4
Shri Bhajan Lal	1655	1575	5.1
Shri Roshan Singh	1880	1555	20.9
Shri Kali Charan	2120	1895	11.9
Shri Ashok Kumar	2070	1720	20.3
Range	1625 - 2120	1370 - 1895	5.1 – 33.3
Standard Error	31.4	30.5	-
Mean	1842	1577	16.8
Sustainability Index	0.80	-	-

Table 3: Residual effect of micro-nutrients applied to groundnut on succeeding crops

Succeeding crop/ Farmer	Grain yield (kg ha^{-1})		% yield increase
	With micro-nutrients (B, Zn) & S	Without micro-nutrients & S	
Wheat			
Shri Ram Dayal	3450	3200	7.8
Shri Atma Ram	2800	2710	3.3
Shri Arun Kumar	3190	2905	9.8
Shri Kallu Sahu	2600	2300	13.0
Mean	3010	2779	8.3
Standard Error	191.1	-	-
Sustainability Index	0.76	-	-
Barley			
Shri Bahadur Singh	3875	3590	7.9
Shri Daya Ram	3360	3010	11.6
Shri Sharda Saran	4450	4120	8.0
Shri Ram Pal Singh	3360	3000	12.0
Shri Arvind Singh	3470	3125	11.0
Shri Puran Rai	3690	3340	10.5
Mean	3701	3364	10.0
Standard Error	171	-	-
Sustainability Index	0.74	-	-
Gram			
Shri Jindel Singh	1680	1570	7.0
Srimati Kasturi Devi	1950	1800	8.3
Shri Vijay Singh	1785	1610	10.9
Shri Om Prakash	1590	1390	14.4
Shri Jagat Ram	1600	1550	3.2
Shri Tula Ram	2090	1910	9.4
Mean	1783	1638	8.8
Standard Error	82.4	-	-
Sustainability Index	0.76	-	-
Mustard			
Shri Bhajan Lal	1350	1230	9.8
Shri Roshan Singh	1210	1090	11.0
Shri Kali Charan	1480	1350	9.7
Shri Ashok Kumar	1500	1405	6.8
Mean	1385	1269	9.2
Standard Error	67.1	-	-
Sustainability Index	0.83	-	

Table 4: Spread of micro-nutrient use and hurdles in adoption (2012)

Particulars	Number of Farmers
Spread of micro-nutrient use (N=80)	
Farmers convinced by the impact of micro-nutrient use	70 (87.5)
Farmers wanted to use micro-nutrient	61(76.3)
Farmers tried to buy micro-nutrient	22 (27.5)
Farmers actually used micro-nutrient	07 (8.7)
Hurdles in adoption of micro-nutrients	
Availability	54 (67.5)
Quality of available micro-nutrient mixture	06 (7.5)
Cost	04 (5.0)
Lack of knowledge and awareness	08 (10.0)
Not effective	08 (10.0)
Figures in parentheses indicates %	

References

Girish Chander, Wani, S.P., Sahrawat, K.L., Pardhasaradhi, G., Rajesh, Rao, C.N., Pal, C.K. and Kamadi, P. 2011. Macro benefits from Zinc, Boron and Sulphur fertilization of rainfed systems in the Semi Arid Zone of India. Resilient Dryland Systems, ICRISAT, Patancheru, A. P., India. (http://www.zinccrops2011.org/presentations/2011_zinccrops2011_chander_abstract.pdf)

Kokate, K.D., Reddy, G.R. and Sudhakar, N. 2012. On-farm validation of nutrients response in different locations of Maharashtra and Andhra Pradesh. *Indian J. Agril. Sci.,* 82 (8): 697–701.

Kokate, K.D., Reddy, G.R., Sudhakar, N., Dattatri, K. and Reddy, K.M. 2010. Plant Nutrient Management for Enhancing Productivity of Crops (Explorations of KVKs), 48 pp. Zonal Project Directorate, Zone-V, Hyderabad.

Palsaniya, D. R., Singh, R., Tewari, R. K., Yadav, R. S. and Dhyani, S. K. 2012a. Integrated watershed management for natural resource conservation and livelihood security in Semi Arid Tropics of India. *Indian J. Agril. Sci.,* 82 (3): 241-247.

Palsaniya, D. R. and Ahlawat, I. P. S. 2009a. Sulphur management in pigeonpea (*Cajanus cajan*)-wheat (*Triticum aestivum*) cropping system. *Indian J. Agron.,* 54(3): 272-277.

Palsaniya, D. R. and Ahlawat, I. P. S. 2007. Crop productivity, quality and nutrient uptake of pigeonpea-wheat cropping system as influenced by sulphur management. *Indian J. Agril. Sci.,* 77(10): 660-663.

Palsaniya, D. R., Tewari, R. K., Singh, R., Yadav, R. S. and Dhyani, S. K. 2010. Farmer - agroforestry land use adoption interface in degraded agroecosystem of Bundelkhand region, India. *Range Managet. Agrofore.,* 31(1): 11-19.

Palsaniya, D. R., Singh, R., Tewari, R. K., Yadav, R. S., Kumar, R. V. and Dhyani, S. K. 2012b. Integrated watershed management for sustainable agricultural production in Semi-Arid Tropics of India. *Indian J. Agron.,* 57 (4): 310-318.

Palsaniya, D. R., Singh, R., Yadav, R. S., Tewari, R. K. and Dhyani, S. K. 2011. Now it is water all the way in Garhkundar - Dabar watershed of drought prone semi-arid Bundelkhand, India. *Curr. Sci.,* 100(9): 1287-1288.

Palsaniya, D. R., Singh, R., Yadav, R. S., Tewari, R. K., Dwivedi, R. P., Kumar, R. V., Venkatesh, A., Kareemulla, K., Bajpai, C. K., Singh, R., Yadav, S. P. S., Chaturvedi, O. P. and Dhyani, S. K. 2009b. Participatory Agro-ecosystem Analysis and Identification of Problems in Garhkundar-Dabar Watershed of Central India. *Indian J. Agrofore.,* 11(1): 91-98.

Palsaniya, D. R., Singh, R., Tewari, R. K., Yadav, R. S., Dwivedi, R. P., Kumar, R. V., Venkatesh, A., Kareemulla, K., Bajpai, C. K., Singh, R., Yadav, S. P. S., Chaturvedi, O. P. and Dhyani, S. K. 2008. Socio economic and livelihood analysis of people in Garhkundar-Dabar watershed of central India. *Indian J. Agrofore.,* 10(1): 65-72.

Rattan, R. K., Patel, K. P., Manjaiah, K. M. and Datta, S. P. 2009. Micro-nutrients in soil, plant, animal and human health. *J. Indian Soc. Soil Sci.,* 57(4):546-558.

Reddy, K. S., Kumar, N., Sharma, A. K., Acharya, C. L. and Dalal, R. C. 2005. Biophysical and sociological impacts of farmyard manure and its potential role in meeting crop nutrient needs: a farmers' survey in Madhya Pradesh, India. *Australian J. Experi. Agric.,* 45: 357-367.

Rego, T. J., Wani, S. P., Saharawat, K. L. and Pardhasaradhi, G. 2005. Macro-benefits from B, Zn and S application in SAT: A step to grey to green revolution in agriculture. Global Theme on Agro ecosystem Report No. 16. ICRISAT, Patencheru, 502324, Andhra Pradesh, India, 24 pp.

Rego, T. J., Sahrawat, K. L., Wani, S. P. and Pardhasaradhi, G. 2007. Widespread deficiencies of Sulfur, Boron, and Zinc in Indian Semi-Arid Tropical soils: On farm crop responses. *J. Plant Nutri.,* 30: 1569-1583.

Samra, J. S. 1997. Status of research on watershed management. CSWCRTI, Dehra Dun.

Saxena, S. K. 1995. India: constraint and opportunities for fertilizer use. Agro-chemicals News in Brief 18 (No. 2).

Shukla, A. K. and Behera, S. K. 2012. Micro-nutrient fertilizers for higher productivity. *Indian J. Ferti.,* 8(4): 100-117.

Shukla, A. K., Behera, S. K., Subba Rao, A. and Singh, A. K. 2012. State wise micro and secondary nutrients recommendations for different crops and cropping systems. Research Bulletin No. 1/2012, IISS, Bhopal, pp. 1-40.

Swarup, A. 1998. Emerging soil fertility management issues for sustainable crop production in irrigated systems: In Long term fertility management through IPNS. ISSS, Bhopal, India, pp. 54-68.

Takkar, P. N. 1996. Micro-nutrient research and sustainable agriculture productivity. *J. Indian Soc. Soil Sci.,* 44:563–581.

Tisdale, S. L., Nelson, W. L. and Beaton, J. D. 1985. Soil fertility and fertilizers. Macmillan Publication, U.S.A.

The Financial Express. 2006. Grave sulphur deficiency in Indian soils. October 5, 2006.

Vasuki, N. 2010. Micro-nutrients management for enhancing crop production - future strategy and requirement. *J. Indian Soc. Soil Sci.,* 58(1): 32-36.

Wani, S. P., Sreedevi, T. K. and Marimuthu, S. 2007. A comprehensive assessment of watershed programs in India. Proceedings of the Review Meeting and Component Workshop, 332 pp, 23–7 July, 2007, ICRISAT, Hyderabad, Andhra Pradesh.

❑❑❑